VIDEO ORGANIZER

BASIC COLLEGE MATHEMATICS
SIXTH EDITION

Elayn Martin-Gay
University of New Orleans

The author and publisher of this book have used their best efforts in preparing this book. These efforts include the development, research, and testing of the theories and programs to determine their effectiveness. The author and publisher make no warranty of any kind, expressed or implied, with regard to these programs or the documentation contained in this book. The author and publisher shall not be liable in any event for incidental or consequential damages in connection with, or arising out of, the furnishing, performance, or use of these programs.

Reproduced by Pearson from electronic files supplied by the author.

Copyright © 2019 by Pearson Education, Inc.
Publishing as Pearson, 501 Boylston Street, Boston, MA 02116.

All rights reserved. No part of this publication may be reproduced, stored in a retrieval system, or transmitted, in any form or by any means, electronic, mechanical, photocopying, recording, or otherwise, without the prior written permission of the publisher. Printed in the United States of America.

4 2021

ISBN-13: 978-0-13-484000-0
ISBN-10: 0-13-484000-3

Table of Contents

Chapter 1 The Whole Numbers 1

Chapter 2 Multiplying and Dividing Fractions 24

Chapter 3 Adding and Subtracting Fractions 38

Chapter 4 Decimals 52

Chapter 5 Ratio and Proportion 66

Chapter 6 Percent 73

Chapter 7 Measurement 87

Chapter 8 Geometry 97

Chapter 9 Statistics and Probability 124

Chapter 10 Signed Numbers 136

Chapter 11 Introduction to Algebra 151

*Answers are available to instructors in the companion MyLab Math course instructor resource section; instructors can make answers available to students at their discretion and upon request.

Basic College Mathematics Sixth Edition, Elayn Martin-Gay Sec. 1.1

Section 1.1 Study Skill Tips for Success in Mathematics

Complete the outline as you view Lecture Video 1.1. Pause ⏸ the video as needed as you fill in all blanks. Circle your answer to each numbered exercise. Then press Play ▶ to continue listening to the video.

Objective A Get ready for this course.

Have a positive attitude!

Check to see that you understand the way that this course is taught.

On the first day of class, bring all the materials.

Objective B Understand some general tips for success.

⏸ Stay _____ !

⏸ Attend all _____ and be on _____ .

Get help as soon as you need it.

⏸ Learn from your _____ .

Turn in assignments on time.

Objective C Know how to use this text.

⏸ Open your _____ and become familiar with it.

⏸ The pencil symbol means a _____ exercise.

⏸ The triangle symbol means the exercise has something to do with _____ .

The ▶ means that the corresponding exercise can be viewed on the DVD lecture series.

Basic College Mathematics Sixth Edition, Elayn Martin-Gay — Sec. 1.1

Section 1.1 Study Skill Tips for Success in Mathematics

Objective D Know how to use text resources.

- The two resources focused on are _____ Resources and _____ Resources.

- DVD Lecture Videos correspond to every section in the book.

- The _____ _____ Prep Videos contain the solutions to all the Chapter Test exercises.

- Student Success Tips are 3 minute reminders.

- The _____ _____ Videos are videos that show the solutions to every exercise in the final practice exam.

- _____ Organizer follows the DVD Lecture Series.

- The _____ Organizer contains tips for note-taking, practice exercises before homework, and references to the text and DVD lecture series.

Objective E Get help as soon as you need it.

- Get help with mathematics as soon as you need it.

Objective F Learn how to prepare for and take an exam.

- When you think you are ready, take a _____ _____.

- Remember the Chapter Test Prep Videos contain the worked out solutions to all end-of-chapter Chapter Test Exercises.

Objective G Develop good time management.

Try filling out the Time Grid at the end of Section 1.1.

Section 1.2 Place Value, Names for Numbers, and Reading Tables

Complete the outline as you view Lecture Video 1.2. Pause ⏸ the video as needed as you fill in all blanks. Circle your answer to each numbered exercise. Then press Play ▶ to continue listening to the video.

Objective A Find the place value of a digit in a whole number.

The digits 0, 1, 2, 3, 4, 5, 7, 8, and 9 can be used to write numbers.

Whole Numbers: 0, 1, 2, 3, 4, 5, ...
Natural Numbers: 1, 2, 3, 4, 5, ...

⏸ The placement of a digit in a number determines its _____ _____.

Determine the place value of the digit 5 in the number

▶ **Work with me.**

1. 657

⏸ **Pause and work.**

2. 5423

▶ Play and check.

Objective B Write a whole number in words and in standard form.

⏸ These whole numbers (657 and 5423) are in written in _____ _____.

⏸ Each group of three digits is called a _____.

Writing a Whole Number in Words
To write a whole number in words, write the number in each period followed by the name of the period. (The ones period is usually not written.) This same procedure can be used to read a whole number.

Section 1.2 Place Value, Names for Numbers, and Reading Tables

Write in words.

Work with me.

3. 26,990

Writing a Whole Number in Standard Form
To write a whole number in standard form, write the number in each period followed by a comma.

Write in standard form.

Work with me.

4. fifty-nine thousand, eight hundred

Objective C Write a whole number in expanded form.

The _____ _____ of a number shows each digit of the number with its place value.

Write in expanded form.

Work with me.

5. 80,774

Objective D Read tables.

Use the table to answer the following questions.

Mountain (State)	Elevation (in feet)
Boott Spur (NH)	5492
Mt. Adams (NH)	5774
Mt. Clay (NH)	5532
Mt Jefferson (NH)	5712
Mt Sam Adams (NH)	5584
Mt. Washington (NH)	6288
Source: U.S. Geological Survey	

Section 1.2 Place Value, Names for Numbers, and Reading Tables

◉ **Pause and work.**

6. Write the elevation of Mt. Clay in standard form and then in words.

◉ Play and check.

◉ **Pause and work.**

7. Write the height of Boot Spur in expanded form.

◉ Play and check.

◉ **Pause and work.**

8. Which mountain is the tallest in New England?

Basic College Mathematics Sixth Edition, Elayn Martin-Gay   Sec. 1.3

Section 1.3 Adding Whole Numbers and Perimeter

Complete the outline as you view Lecture Video 1.3. Pause ⏸ the video as needed as you fill in all blanks. Circle your answer to each numbered exercise. Then press Play ▶ to continue listening to the video.

Objective A Add whole numbers.

If needed, study sums of 1-digit numbers.

Addition Property of 0
The sum of 0 and any number is that number.
$$7 + 0 = 7$$
$$0 + 7 = 7$$

⏸ Commutative properties have to do with _____.

Commutative Property of Addition
Changing the order of two addends does not change their sum.
$$2 + 3 = 5$$
and
$$3 + 2 = 5$$

⏸ Associative properties have to do with _____.

Associative Property of Addition
Changing the grouping of addends does not change their sum.
$$3 + (5 + 7) = 3 + 12 = 15$$
and
$$(3 + 5) + 7 = 8 + 7 = 15$$

Add the whole numbers.

▶ **Work with me.**

1. 5267
   + 132
   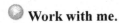

Section 1.3 Adding Whole Numbers and Perimeter

Work with me.

2.   8
       9
       2
       5
    + 1
    ―――

When adding, line up place value digits.

Pause and work.

3.   24
      9006
       489
    + 2407
    ――――

Play and check.

Objective B   Find the perimeter of a polygon.

The _____ of a polygon is the distance around the polygon.

Work with me.

4. Find the perimeter of the triangle.

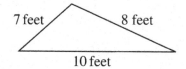

7 feet   8 feet

10 feet

Objective C   Solve problems by adding whole numbers.

Sum means _____.

Basic College Mathematics Sixth Edition, Elayn Martin-Gay

Section 1.3 Adding Whole Numbers and Perimeter

Work with me.

5. Find the sum of 297 and 1796.

Pause and work.

6. The highest point in South Carolina is Sassafras Mountain at 3560 feet above sea level. The highest point in North Carolina is Mt. Mitchell, whose peak is 3124 feet increased by the height of Sassafras Mountain. Find the height of Mt. Mitchell.

Play and check.

Is means _____.

Increased by means _____.

Basic College Mathematics Sixth Edition, Elayn Martin-Gay   Sec. 1.4

Section 1.4 Subtracting Whole Numbers

Complete the outline as you view Lecture Video 1.4. Pause ⏸ the video as needed as you fill in all blanks. Circle your answer to each numbered exercise. Then press Play ▶ to continue listening to the video.

Objective A Subtract whole numbers.

Subtraction Properties of 0
The difference of any number and that same number is 0.
$$11 - 11 = 0$$

The difference of any number and 0 is that same number.
$$45 - 0 = 45$$

Like addition, line up place value digits when subtracting.

Perform the indicated operation.

▶ **Work with me.**

1. 749
    −149
    _____

Check subtraction by addition.

Perform the indicated operation.

⏸ **Pause and work.**

2.   62
    − 37
    _____

▶ Play and check.

▶ **Work with me.**

3.   51,111
    −19,898
    _____

9

Basic College Mathematics Sixth Edition, Elayn Martin-Gay   Sec. 1.4

Section 1.4 Subtracting Whole Numbers

Objective B Solve problems by subtracting whole numbers.

With subtraction, order matters.

Work with me.

4. Subtract 5 from 9.

Difference means _____.

Pause and work.

5. Find the difference of 41 and 21.

Play and check.

Work with me.

6. The Oroville Dam, on the Feather River, is the tallest dam in the United States at 754 feet. The Hoover Dam, on the Colorado River, is 726 feet high. How much taller is the Oroville Dam than the Hoover Dam?

Section 1.5 Rounding and Estimating

Complete the outline as you view Lecture Video 1.5. Pause the video as needed as you fill in all blanks. Circle your answer to each numbered exercise. Then press Play to continue listening to the video.

Objective A Round whole numbers.

⏸ _____ a whole number means approximating it.

⏸ Write the steps used to round whole numbers to a given place value.

Rounding Whole Numbers to a Given Place Value
Step 1:

Step 2:

Step 3:

▶ **Work with me.**

1. Round 635 to the nearest ten.

⏸ **Pause and work.**

2. Round 36,499 to the nearest thousand.

▶ Play and check.

Section 1.5 Rounding and Estimating

Objective B Use rounding to estimate sums and differences.

Work with me.

3. Estimate by rounding to the nearest ten.

   ```
    39
    45
    22
   +17
   ```

Pause and work.

4. Estimate by rounding to the nearest hundred.

   ```
    1774
   −1492
   ```

Play and check.

Objective C Solve problems by estimating.

Solve the word problem.

Work with me.

5. The peak of Mt. McKinley, in Alaska, is 20,320 feet above sea level. The top of Mt. Rainier, in Washington, is 14,410 feet above sea level. Round each height to the nearest thousand to estimate the difference in elevation of these two peaks.

Basic College Mathematics Sixth Edition, Elayn Martin-Gay   Sec. 1.6

Section 1.6 Multiplying Whole Numbers and Area

Complete the outline as you view Lecture Video 1.6. Pause the video as needed as you fill in all blanks. Circle your answer to each numbered exercise. Then press Play to continue listening to the video.

Objective A Use the properties of multiplication.

The result of an addition problem is called the ____ or total.

The result of a multiplication problem is called the _____.

Multiplication Property of 0
The product of 0 and any number is 0.
$$5 \cdot 0 = 0$$
and
$$0 \cdot 8 = 0$$

Multiplication Property of 1
The product of 1 and any number is that same number.
$$1 \cdot 9 = 9$$
and
$$6 \cdot 1 = 6$$

Perform the indicated operation.

Work with me.

1. $1 \cdot 24$

Pause and work.

2. $0 \cdot 19$

Play and check.

Commutative Property of Multiplication
Changing the order of two factors does not change their product.
$$9 \cdot 2 = 18$$
and
$$2 \cdot 9 = 18$$

13

Section 1.6 Multiplying Whole Numbers and Area

Associative Property of Multiplication
Changing the grouping of factors does not change their product.
$(2 \cdot 3) \cdot 4 = 2 \cdot (3 \cdot 4)$

Distributive Property
Multiplication distributes over addition.

$2(3 + 4) = 2 \cdot 3 + 2 \cdot 4$

Rewrite the expression using the Distributive Property.

▶ **Work with me.**

3. $20(14 + 6)$

Objective B Multiply whole numbers.

Perform the indicated operation.

▶ **Work with me.**

4. 277
   × 6

⏸ **Pause and work.**

5. 8649
   ×274

▶ **Play and check.**

Objective C Multiply whole numbers ending in zero(s).

Perform the indicated operation.

▶ **Work with me**

6. 8×10
   8×100
   8×1000

Section 1.6 Multiplying Whole Numbers and Area

Pause and work.

7. 50 · 900

Play and check.

Objective D Find the area of a rectangle.

_____ measures the surface of a region.

_____ measures the distance around a polygon.

Work with me.

8. Find the area of the rectangle.

   9 meters

   7 meters

Area is measured in _____ units.

Objective E Solve problems by multiplying whole numbers.

Solve the word problem.

Work with me.

9. One tablespoon of olive oil contains 125 calories. How many calories are in 3 tablespoons of olive oil?

Basic College Mathematics Sixth Edition, Elayn Martin-Gay   Sec. 1.7

Section 1.7 Dividing Whole Numbers

Complete the outline as you view Lecture Video 1.7. Pause ⏸ the video as needed as you fill in all blanks. Circle your answer to each numbered exercise. Then press Play ▶ to continue listening to the video.

Objective A Divide whole numbers.

⏸ _____ is the process of separating a quantity into equal parts.

Perform the indicated operation.

▶ **Work with me.**

1. $36 \div 3$

⏸ **Pause and work.**

2. $31 \div 1$

▶ **Play and check.**

▶ **Work with me.**

3. $25 \div 1$

⏸ **Pause and work.**

4. $\dfrac{18}{18}$

▶ **Play and check.**

Section 1.7 Dividing Whole Numbers

🔘 Work with me.

5. $12 \div 12$

Division Properties of 1
The quotient of any number (except 0) and that same number is 1. The quotient of any number and 1 is that same number.

Perform the indicated operation.

🔘 Work with me.

6. $0 \div 14$

⏸ Pause and work.

7. $\dfrac{0}{5}$

🔘 Play and check.

🔘 Work with me.

8. $26 \div 0$

Division Properties of 0
The quotient of 0 and any number (except 0) is 0. The quotient of any number and 0 is undefined.

Objective B Perform long division.

Perform the indicated operation.

🔘 Work with me.

9. $55\overline{)715}$

Basic College Mathematics Sixth Edition, Elayn Martin-Gay   Sec. 1.7

Section 1.7 Dividing Whole Numbers

⏸ **Pause and work.**

10. $20{,}619 \div 102$

🔘 Play and check.

⏸ The remainder is ____.

To check: $202 \cdot 102 + 15 - 20{,}619$ (the dividend)

Objective C Solve problems that require dividing by whole numbers.

🔘 **Work with me.**

11. An 18-hole golf course is 5580 yards long. If the distance to each hole is the same, find the distance between holes.

To check: $310 \cdot 18 - 5580$ (the dividend)

Objective D Find the average of a list of numbers.

Average of a List of Numbers

$$\text{average} = \frac{\text{sum of numbers}}{\text{number of numbers}}$$

Find the average of the list of numbers.

🔘 **Work with me.**

12. 86, 79, 81, 69, 80

Basic College Mathematics Sixth Edition, Elayn Martin-Gay Sec. 1.8

Section 1.8 An Introduction to Problem Solving

Complete the outline as you view Lecture Video 1.8. Pause ⏸ the video as needed as you fill in all blanks. Circle your answer to each numbered exercise. Then press Play ▶ to continue listening to the video.

Objective A Solve problems by adding, subtracting, multiplying, or dividing whole numbers.

Addition (+)	Subtraction (−)	Multiplication (·)	Division (÷)	Equality (=)
sum	difference	product	quotient	equals
plus	minus	times	divide	is equal to
added to	subtract	multiply	shared	is/was
more than	less than	multiply by	equally	yields
increased by	decreased by	of	among	
total	less	double/triple	per	
			divided by	
			divided into	

⏸ Write the steps used to solve problems.

Problem-Solving Steps

Step 1:

Step 2:

Step 3:

Step 4:

⏸ _____ means division.

⏸ Is means _____.

19

Basic College Mathematics Sixth Edition, Elayn Martin-Gay   Sec. 1.8

Section 1.8 An Introduction to Problem Solving

🔵 **Work with me.**

1. What is the quotient of 1185 and 5?

⏸ Total means _____.

⏸ **Pause and work.**

2. What is the total of 35 and 7?

🔵 **Play and check.**

Solve the application problem.

🔵 **Work with me.**

3. The Verrazano Narrows Bridge is the longest bridge in New York, measuring 4260 feet. The George Washington Bridge, also in New York, is 760 feet shorter than the Verrazano Narrows Bridge. Find the length of the George Washington Bridge.

Objective B Solve problems that require more than one operation.

🔵 **Work with me.**

4. Find the total cost of 3 sweaters at $38 each and 5 shirts at $25 each.

Section 1.9 Exponents, Square Roots, and Order of Operations

Complete the outline as you view Lecture Video 1.9. Pause the video as needed as you fill in all blanks. Circle your answer to each numbered exercise. Then press Play to continue listening to the video.

Objective A Write repeated factors using exponential notation.

A shorthand notation for repeated multiplication of the same factor is called _____.

Rewrite each expression using exponential notation.

Work with me.

1. $12 \cdot 12 \cdot 12$

Pause and work.

2. $6 \cdot 6 \cdot 5 \cdot 5 \cdot 5$

Play and check.

Usually, an exponent of 1 is not written.

Objective B Evaluate expressions containing exponents.

Evaluate each expression.

Work with me.

3. 5^3

Pause and work.

4. 7^1

Play and check.

Work with me.

5. 10^2

Basic College Mathematics Sixth Edition, Elayn Martin-Gay   Sec. 1.9

Section 1.9 Exponents, Square Roots, and Order of Operations

Objective C Evaluate the square root of a perfect square.

A _____ of a number is one of two identical factors of the number.

$\sqrt{}$ is a _____ .

Evaluate each expression.

Work with me.

6. $\sqrt{9}$

Pause and work.

7. $\sqrt{64}$

Play and check.

Objective D Use the order of operations.

Write the steps used in order of operations.

Order of Operations
Step 1:
Step 2:
Step 3:
Step 4:

22

Section 1.9 Exponents, Square Roots, and Order of Operations

Evaluate each expression.

Work with me.

8. $15 + 3 \cdot 2$

Pause and work.

9. $14 \div 7 \cdot 2 + 3$

Play and check.

Work with me.

10. $\dfrac{7(9-6)+3}{3^2 - 3}$

Pause and work.

11. $(7 \cdot 5) + [9 \div (3 \div 3)]$

Play and check.

Objective E Find the area of a square.

Find the area and perimeter of the square.

Work with me.

12.

7 meters

Basic College Mathematics Sixth Edition, Elayn Martin-Gay   Sec. 2.1

Section 2.1 Introduction to Fractions and Mixed Numbers

Complete the outline as you view Lecture Video 2.1. Pause ⏸ the video as needed as you fill in all blanks. Circle your answer to each numbered exercise. Then press Play ▶ to continue listening to the video.

Objective A Identify the numerator and the denominator of a fraction and review division properties of 0 and 1.

⏸ To refer to part of a whole, we can use _____.

⏸ The number on top is the _____.

⏸ The number on bottom is the _____.

Identify the parts of the fraction.

▶ **Work with me.**

1. $\dfrac{1}{2}$

⏸ The _____ _____ means division.

⏸ **Pause and work.**

2. $\dfrac{10}{3}$

▶ Play and check.

Simplify the fraction.

▶ **Work with me.**

3. $\dfrac{21}{21}$

24

Section 2.1 Introduction to Fractions and Mixed Numbers

⏸ **Pause and work.**

4. $\dfrac{13}{1}$

▶ Play and check.

▶ **Work with me.**

5. $\dfrac{0}{20}$

▶ **Work with me.**

6. $\dfrac{5}{0}$

Let *n* be any whole number except 0.

$\dfrac{n}{n} = 1 \qquad \dfrac{0}{n} = 0$

$\dfrac{n}{1} = n \qquad \dfrac{n}{0}$ is undefined.

Objective B Write a fraction to represent parts of figures or real-life data.

▶ **Work with me.**

7. What portion of this figure is shaded?

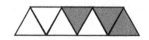

Section 2.1 Introduction to Fractions and Mixed Numbers

🔵 **Work with me.**

8. Marbles: 50 red or blue; 21 blue

    a. What fraction of the marbles are blue?

    b. How many marbles are red?

    c. What fraction of the marbles are red?

Objective C Identify proper fractions, improper fractions, and mixed numbers.

A proper fraction is a fraction whose numerator is less than its denominator. Proper fractions are less than 1.

An improper fraction is a fraction whose numerator is greater than or equal to its denominator. Improper fractions are greater than or equal to 1.

A mixed number contains a whole number and a fraction. Mixed numbers are greater than 1.

Determine whether the fraction is proper or improper.

🔵 **Work with me.**

9. $\dfrac{1}{2}$

⏸ **Pause and work.**

10. $\dfrac{10}{3}$

🔵 **Play and check.**

Section 2.1 Introduction to Fractions and Mixed Numbers

◉ **Work with me.**

11. Consider the group of figures. Represent the shading of this group as an improper fraction and as a mixed number.

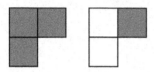

Objective D Write mixed numbers as improper fractions.

⏸ A _____ _____ has a whole number part and a fraction part.

⏸ Write the steps used to write a mixed number as an improper fraction.

Writing a Mixed Number as an Improper Fraction
Step 1:
Step 2:
Step 3:

Write the mixed number as an improper fraction.

◉ **Work with me.**

12. $2\dfrac{1}{3}$

⏸ **Pause and work.**

13. $3\dfrac{3}{5}$

◉ **Play and check.**

Basic College Mathematics Sixth Edition, Elayn Martin-Gay   Sec. 2.1

Section 2.1 Introduction to Fractions and Mixed Numbers

⏸ **Pause and work.**

14. $9\dfrac{7}{20}$

▶ Play and check.

Objective E Write improper fractions as mixed numbers or whole numbers.

⏸ Write the steps used to write an improper fraction as a mixed number or a whole number.

Writing an Improper Fraction as a Mixed Number or a Whole Number
Step 1:
Step 2:

Write the improper fraction as an equivalent mixed number or whole number.

▶ **Work with me.**

15. $\dfrac{17}{5}$

⏸ **Pause and work.**

16. $\dfrac{37}{8}$

▶ Play and check.

Basic College Mathematics Sixth Edition, Elayn Martin-Gay   Sec. 2.2

Section 2.2 Factors and Prime Factorization

Complete the outline as you view Lecture Video 2.2. Pause ⏸ the video as needed as you fill in all blanks. Circle your answer to each numbered exercise. Then press Play ▶ to continue listening to the video.

Objective A Find the factors of a number.

⏸ 2·5 is a _____ of 10.

Find the factors of the number.

▶ **Work with me.**

1. 25

⏸ **Pause and work.**

2. 12

▶ Play and check.

Objective B Identify prime and composite numbers.

Prime Numbers
A prime number is a natural number that has exactly two different factors, 1 and itself.

Composite Numbers
A composite number is any natural number, other than 1, that is not prime.

The natural number 1 is neither prime nor composite.

Determine whether the number is prime or composite.

▶ **Work with me.**

3. 4

Section 2.2 Factors and Prime Factorization

⏸ **Pause and work.**

4. 10

▶ Play and check.

▶ **Work with me.**

5. 67

Objective C Find the prime factorization of a number.

Prime Factorization
The prime factorization of a number is the factorization in which all the factors are prime numbers.

Find the prime factorization of the number.

▶ **Work with me.**

6. 15

⏸ **Pause and work.**

7. 36

▶ Play and check.

▶ **Work with me.**

8. 240

Basic College Mathematics Sixth Edition, Elayn Martin-Gay   Sec. 2.3

Section 2.3 Simplest Form of a Fraction

Complete the outline as you view Lecture Video 2.3. Pause ⏸ the video as needed as you fill in all blanks. Circle your answer to each numbered exercise. Then press Play ▶ to continue listening to the video.

Objective A **Write a fraction in simplest form or lowest terms.**

⏸ Fractions that represent the same portion of a whole are _____ _____.

Simplest Form of a Fraction
A fraction is written in simplest form or lowest terms when the numerator and the denominator have no common factors other than 1.

Simplify.

▶ **Work with me.**

1. $\dfrac{14}{16}$

⏸ **Pause and work.**

2. $\dfrac{24}{40}$

▶ Play and check.

⏸ **Pause and work.**

3. $\dfrac{70}{196}$

▶ Play and check.

Objective B **Determine whether two fractions are equivalent.**

Two fractions are equivalent if they simplify to the same fraction.

Section 2.3 Simplest Form of a Fraction

Cross Products

$$\frac{16}{40} = \frac{10}{25}$$

Cross Products: $25 \cdot 16, \ 40 \cdot 10$

If cross products are equal, the fractions are equivalent. If cross products are not equal, the fractions are not equivalent.

Determine whether the fractions are equivalent or not.

▶ **Work with me.**

4. $\dfrac{7}{11}$ and $\dfrac{5}{8}$

⏸ **Pause and work.**

5. $\dfrac{3}{9}$ and $\dfrac{6}{18}$

▶ **Play and check.**

Objective C Solve problems by writing fractions in simplest form.

▶ **Work with me.**

6. There are 5280 feet in a mile. What fraction of a mile is represented by 2640 feet?

▶ **Work with me.**

7. The outer wall of the Pentagon is 24 inches wide. Ten inches is concrete, 8 inches is brick, and 6 inches is limestone. What fraction of the wall is concrete?

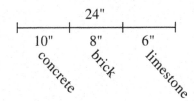

Section 2.4 Multiplying Fractions and Mixed Numbers

Complete the outline as you view Lecture Video 2.4. Pause the video as needed as you fill in all blanks. Circle your answer to each numbered exercise. Then press Play to continue listening to the video.

Objective A Multiply fractions.

Multiplying Fractions
To multiply two fractions, multiply the numerators and multiply the denominators. If a, b, c, and d represent positive whole numbers, we have
$$\frac{a}{b} \cdot \frac{c}{d} = \frac{a \cdot c}{b \cdot d}.$$

Multiply the fractions.

Work with me.

1. $\dfrac{6}{5} \cdot \dfrac{1}{7}$

Pause and work.

2. $\dfrac{2}{7} \cdot \dfrac{5}{8}$

Play and check.

Work with me.

3. $\dfrac{11}{20} \cdot \dfrac{1}{7} \cdot \dfrac{5}{22}$

Objective B Multiply Fractions and Mixed Numbers or Whole Numbers.

Multiplying Fractions and Mixed Numbers or Whole Numbers
To multiply with mixed numbers or whole numbers, first write any mixed or whole numbers as fractions and then multiply as usual.

Section 2.4 Multiplying Fractions and Mixed Numbers

Perform the multiplication.

▶ **Work with me.**

4. $\dfrac{5}{8} \cdot 4$

▶ **Work with me.**

5. $2\dfrac{1}{5} \cdot 3\dfrac{1}{2}$

Objective C Solve problems by multiplying fractions.

▶ **Work with me.**

6. The Oregon National Historic Trail is 2170 miles long. It begins in Independence, Missouri, and ends in Oregon City, Oregon. Manfred Coulon has hiked $\dfrac{2}{5}$ of the trail before. How many miles has he hiked?

⏸ _____ usually translates to _____.

▶ **Work with me.**

7. The radius of a circle is one-half of its diameter. If the diameter of a circle is $\dfrac{3}{8}$ of an inch, what is its radius?

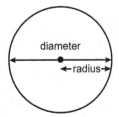

Section 2.5 Dividing Fractions and Mixed Numbers

Complete the outline as you view Lecture Video 2.5. Pause the video as needed as you fill in all blanks. Circle your answer to each numbered exercise. Then press Play to continue listening to the video.

Objective A Find the reciprocal of a fraction.

Reciprocal of a Fraction
Two numbers are reciprocals of each other if their product is 1. The reciprocal of the fraction
$$\frac{a}{b} \text{ is } \frac{b}{a}$$
because $\frac{a}{b} \cdot \frac{b}{a} = \frac{a \cdot b}{b \cdot a} = 1.$

Finding the Reciprocal of a Fraction
To find the reciprocal of a fraction, interchange its numerator and denominator.

Find the reciprocal.

Work with me.

1. $\frac{4}{7}$

Work with me.

2. 15

Objective B Divide fractions.

Dividing Fractions
To divide two fractions, multiply the first fraction by the reciprocal of the second fraction.
If a, b, c, and d represent numbers, and b, c, and d are not 0, then
$$\frac{a}{b} \div \frac{c}{d} = \frac{a}{b} \cdot \frac{d}{c} = \frac{a \cdot d}{b \cdot c}.$$

Basic College Mathematics Sixth Edition, Elayn Martin-Gay Sec. 2.5

Section 2.5 Dividing Fractions and Mixed Numbers

Divide the fractions.

Work with me.

3. $\dfrac{2}{3} \div \dfrac{5}{6}$

Pause and work.

4. $\dfrac{1}{10} \div \dfrac{10}{1}$

Play and check.

Work with me.

5. $\dfrac{3}{25} \div \dfrac{27}{40}$

Work with me.

6. $\dfrac{8}{13} \div 0$

> 0 has _____ reciprocal.

Objective C Divide fractions and mixed numbers or whole numbers.

Dividing Fractions and Mixed Numbers or Whole Numbers
To divide with a mixed number or a whole number, first write the mixed or whole number as a fraction and then divide as usual.

Divide the fractions and mixed numbers or whole numbers.

Work with me.

7. $\dfrac{2}{3} \div 4$

Section 2.5 Dividing Fractions and Mixed Numbers

Work with me.

8. $3\dfrac{3}{7} \div 3\dfrac{1}{3}$

Objective D Solve problems by dividing fractions.

Work with me.

9. A patient is to take $3\dfrac{1}{3}$ tablespoons of medicine per day in 4 equally divided doses. How much medicine is to be taken in each dose?

Basic College Mathematics Sixth Edition, Elayn Martin-Gay   Sec. 3.1

Section 3.1 Adding and Subtracting Like Fractions

Complete the outline as you view Lecture Video 3.1. Pause ⏸ the video as needed as you fill in all blanks. Circle your answer to each numbered exercise. Then press Play ▶ to continue listening to the video.

Objective A Add like fractions.

> **Adding Like Fractions (Fractions with the Same Denominator)**
> To add like fractions, add the numerators and write the sum over the common denominator.
> If a, b, and c represent nonzero whole numbers, we have
> $$\frac{a}{c} + \frac{b}{c} = \frac{a+b}{c}.$$

Add.

▶ **Work with me.**

1. $\dfrac{1}{7} + \dfrac{2}{7}$

⏸ **Pause and work.**

2. $\dfrac{2}{9} + \dfrac{4}{9}$

▶ **Play and check.**

⏸ **Pause and work.**

3. $\dfrac{4}{13} + \dfrac{2}{13} + \dfrac{1}{13}$

▶ **Play and check.**

Objective B Subtract like fractions.

> **Subtracting Like Fractions (Fractions with the Same Denominator)**
> To subtract like fractions, subtract the numerators and write the difference over the common denominator.
> If a, b, and c represent nonzero whole numbers, we have
> $$\frac{a}{c} - \frac{b}{c} = \frac{a-b}{c}.$$

Section 3.1 Adding and Subtracting Like Fractions

Subtract.

🌐 **Work with me.**

4. $\dfrac{10}{11} - \dfrac{4}{11}$

⏸ **Pause and work.**

5. $\dfrac{7}{8} - \dfrac{1}{8}$

🌐 **Play and check.**

Objective C Solve problems by adding or subtracting like fractions.

⏸ _____ means distance around.

🌐 **Work with me.**

6. Find the perimeter.

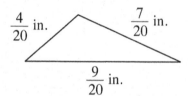

Do not forget to attach units.

⏸ **Pause and work.**

7. Emil Vasquez, a bodybuilder, worked out $\dfrac{7}{8}$ of an hour one morning before school and $\dfrac{5}{8}$ of an hour that evening. How long did he work out that day?

🌐 **Play and check.**

Do not forget to attach units.

Basic College Mathematics Sixth Edition, Elayn Martin-Gay   Sec. 3.2

Section 3.2 Least Common Multiple

Complete the outline as you view Lecture Video 3.2. Pause ⏸ the video as needed as you fill in all blanks. Circle your answer to each numbered exercise. Then press Play ▶ to continue listening to the video.

Objective A **Find the least common multiple (LCM) using multiples.**

⏸ The _____ of a number is the product of that number and natural numbers.

⏸ The _____ numbers are 1, 2, 3, 4, ...

⏸ The smallest common multiple is the _____ _____ _____ (LCM).

⏸ Write the steps used to find the LCM of a list of numbers using multiples of the largest number.

Method 1: Find the LCM of a List of Numbers Using Multiples of the Largest Number.

Step 1:

Step 2:

Find the LCM.

▶ **Work with me.**

1. 9, 15

Section 3.2 Least Common Multiple

Objective B Find the LCM using prime factorization.

Write the steps used to find the LCM of a list of numbers using prime factorization.

Method 2: Finding the LCM of a List of Numbers Using Prime Factorization.
Step 1:
Step 2:
Step 3:

Find the LCM.

Work with me.

2. 8, 24

Pause and work.

3. 25, 15, 6

Play and check.

Pause and work.

4. 30, 36, 50

Play and check.

Basic College Mathematics Sixth Edition, Elayn Martin-Gay — Sec. 3.2

Section 3.2 Least Common Multiple

Objective C Write equivalent fractions.

To write an equivalent fraction,
$$\frac{a}{b} = \frac{a}{b} \cdot \frac{c}{c} = \frac{a \cdot c}{b \cdot c}$$
where a, b, and c are nonzero numbers.

Write an equivalent fraction with the given denominator.

🔵 **Work with me.**

5. $\dfrac{4}{7} = \dfrac{}{35}$

Multiplying by 1 does not change the value of the fraction.

⏸ **Pause and work.**

6. $\dfrac{4}{9} = \dfrac{}{81}$

🔵 **Play and check.**

Section 3.3 Adding and Subtracting Unlike Fractions

Complete the outline as you view Lecture Video 3.3. Pause the video as needed as you fill in all blanks. Circle your answer to each numbered exercise. Then press Play to continue listening to the video.

Objective A Add unlike fractions.

Write the steps used to add or subtract unlike fractions.

Adding or Subtracting Unlike Fractions
Step 1:
Step 2:
Step 3:
Step 4:

Add.

Work with me.

1. $\dfrac{2}{11} + \dfrac{2}{33}$

Pause and work.

2. $\dfrac{7}{15} + \dfrac{5}{12}$

Play and check.

Objective B Subtract unlike fractions.

Subtract.

Section 3.3 Adding and Subtracting Unlike Fractions

🔵 **Work with me.**

3. $\dfrac{5}{6} - \dfrac{3}{7}$

⏸️ **Pause and work.**

4. $\dfrac{11}{35} - \dfrac{2}{7}$

▶️ **Play and check.**

Objective C Solve problems by adding or subtracting unlike fractions.

⏸️ _____ means distance around the figure.

🔵 **Work with me.**

5. Find the perimeter.

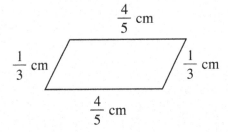

Top: $\dfrac{4}{5}$ cm; Left: $\dfrac{1}{3}$ cm; Right: $\dfrac{1}{3}$ cm; Bottom: $\dfrac{4}{5}$ cm

Section 3.4 Adding and Subtracting Mixed Numbers

Complete the outline as you view Lecture Video 3.4. Pause the video as needed as you fill in all blanks. Circle your answer to each numbered exercise. Then press Play to continue listening to the video.

Objective A Add mixed numbers.

To add or subtract mixed numbers, keep the mixed-number format.

Adding or Subtracting Mixed Numbers
To add or subtract mixed numbers, add or subtract the fraction parts and then add or subtract the whole number parts.

Add.

Work with me.

1. $\begin{array}{r} 10\frac{3}{14} \\ +\ 3\frac{4}{7} \\ \hline \end{array}$

Pause and work.

2. $\begin{array}{r} 1\frac{5}{6} \\ +5\frac{3}{8} \\ \hline \end{array}$

Play and check.

Section 3.4 Adding and Subtracting Mixed Numbers

Objective B Subtract mixed numbers.

Subtract.

🔘 **Work with me.**

3. 6
    $-2\ \frac{4}{9}$
    ─────

To subtract mixed numbers, it is sometimes necessary to borrow.

⑪ To check subtraction, use _____.

Objective C Solve problems by adding or subtracting mixed numbers.

Solve the application.

🔘 **Work with me.**

4. If Tucson's average annual rainfall is $11\frac{1}{4}$ inches and Yuma's is $3\frac{3}{5}$ inches, how much more rain, on average, does Tucson get than Yuma?

Basic College Mathematics Sixth Edition, Elayn Martin-Gay   Sec. 3.5

Section 3.5 Order, Exponents, and the Order of Operations

Complete the outline as you view Lecture Video 3.5. Pause the video as needed as you fill in all blanks. Circle your answer to each numbered exercise. Then press Play to continue listening to the video.

Objective A Compare fractions.

< means is less than.
> means is greater than.

Write the steps used to compare fractions.

Comparing Fractions

To determine which of two fractions is greater,

Step 1:

Step 2:

Insert the symbol < or > to make a true statement.

Work with me.

1. $\dfrac{3}{3}$ $\dfrac{5}{3}$

Pause and work.

2. $\dfrac{3}{5}$ $\dfrac{9}{14}$

Play and check.

Objective B Evaluate fractions raised to powers.

Simplify.

Work with me.

3. $\left(\dfrac{2}{5}\right)^3$

Basic College Mathematics Sixth Edition, Elayn Martin-Gay   Sec. 3.5

Section 3.5 Order, Exponents, and the Order of Operations

Objective C Review operations on fractions.

Review of Operations on Fractions

Multiply	Multiply the numerators and multiply the denominators.
Divide	Multiply the first fraction by the reciprocal of the second fraction.
Add or Subtract	1. Write each fraction as an equivalent fraction whose denominator is the LCD. 2. Add or subtract numerators and write the result over the common denominator.

Operations on Mixed Numbers
To multiply or divide mixed numbers, write each mixed number as an equivalent improper fraction.
To add or subtract mixed numbers, leave the numbers in mixed-number format.

Add.

Work with me.

4. $4\frac{2}{9} + 5\frac{9}{11}$

Objective D Use the order of operations.

Write the steps used in the order of operations.

Order of Operations

Step 1:

Step 2:

Step 3:

Step 4:

Section 3.5 Order, Exponents, and the Order of Operations

Simplify.

Work with me.

5. $\dfrac{1}{5} + \dfrac{1}{3} \cdot \dfrac{1}{4}$

Pause and work.

6. $\left(\dfrac{2}{3} - \dfrac{5}{9}\right)^2$

Play and check.

Basic College Mathematics Sixth Edition, Elayn Martin-Gay   Sec. 3.6

Section 3.6 Fractions and Problem Solving

Complete the outline as you view Lecture Video 3.6. Pause ⏸ the video as needed as you fill in all blanks. Circle your answer to each numbered exercise. Then press Play ▶ to continue listening to the video.

Objective A Solve problems by performing operations on fractions or mixed numbers.

Addition (+)	Subtraction (−)	Multiplication (·)	Division (÷)	Equality (=)
sum	difference	product	quotient	equals
plus	minus	times	divide	is equal to
added to	subtract	multiply	shared equally	is/was
more than	less than	multiply by	per	yields
increased by	decreased by	of	divided by	
total	less	double/triple	divided into	

⏸ Write the steps used to solve problems.

Problem-Solving Steps

Step 1:

Step 2:

Step 3:

Step 4:

Solve the application problems.

▶ **Work with me.**

1. The life expectancy of a circulating coin is 30 years. The life expectancy of a circulating dollar bill is only $\frac{1}{20}$ as long. Find the life expectancy of the circulating dollar bill.

Section 3.6 Fractions and Problem Solving

Pause and work.

2. Suppose that the cross section of a piece of pipe looks like the diagram shown below. Find the total outer diameter.

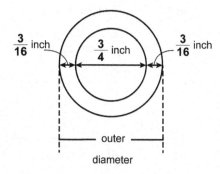

Play and check.

Basic College Mathematics Sixth Edition, Elayn Martin-Gay   Sec. 4.1

Section 4.1 Introduction to Decimals

Complete the outline as you view Lecture Video 4.1. Pause ⏸ the video as needed to fill in all blanks. Circle your answer to each numbered exercise. Then press Play ▶ to continue listening to the video.

Objective A Know the meaning of place value for a decimal number, and write decimals in words.

⏸ Numbers written in decimal notation are called _____ _____, or simply _____.

⏸ Write the steps used to write (or read) a decimal in words.

Writing (or Reading) a Decimal in Words

Step 1:

Step 2:

Step 3:

Write the decimal in words.

▶ **Work with me.**

1. 16.23

⏸ **Pause and work.**

2. 167.009

▶ Play and check.

52

Section 4.1 Introduction to Decimals

Objective B Write decimals in standard form.

Write in standard form.

Work with me.

3. Nine and eight hundredths

Pause and work.

4. Forty-six ten-thousandths

Play and check.

Objective C Write decimals as fractions.

By reading a decimal number correctly, one can write it correctly as a fraction (or mixed number).

Write as a fraction or mixed number.

Work with me.

5. 0.27

Pause and work.

6. 7.008

Play and check.

Objective D Write fractions as decimals.

If the denominator is a power of 10, we can write fractions as decimals by reading them correctly.

Section 4.1 Introduction to Decimals

Simplify each fraction by dividing.

Work with me.

7. $\dfrac{6}{10}$

Pause and work.

8. $\dfrac{45}{100}$

Play and check.

Pause and work.

9. $\dfrac{28}{1000}$

Play and check.

Basic College Mathematics Sixth Edition, Elayn Martin-Gay   Sec. 4.2

Section 4.2 Order and Rounding

Complete the outline as you view Lecture Video 4.2. Pause ⏸ the video as needed as you fill in all blanks. Circle your answer to each numbered exercise. Then press Play ▶ to continue listening to the video.

Objective A Compare decimals.

Comparing Decimals
Compare digits in the same places from left to right. When two digits are not equal, the number with the larger digit is the larger decimal. If necessary, insert 0s after the last digit to the right of the decimal point to continue comparing.

Compare the decimals.

▶ **Work with me.**

1. 0.57 ☐ 0.54

⏸ **Pause and work.**

2. 167.908 ☐ 167.98

▶ **Play and check.**

Objective B Round a decimal number to a given place value.

⏸ Write the steps used to round decimals to a place value to the right of the decimal point.

Rounding Decimals to a Place Value to the Right of the Decimal Point

Step 1:

Step 2:

Section 4.2 Order and Rounding

Round the decimal.

▶ **Work with me.**

3. 0.234 to the nearest hundredth

▶ **Work with me.**

4. 98,207.23 to the nearest ten

⏸ **Pause and work.**

5. 0.1295 to the nearest thousandth

⏸ **Pause and work.**

6. $26.95 to the nearest dollar

▶ Play and check.

▶ Play and check.

⏸ **Pause and work.**

7. The length of a day on Mars is 24.6229 hours. Round this figure to the nearest thousandth.

▶ Play and check.

Section 4.3 Adding and Subtracting Decimals

Complete the outline as you view Lecture Video 4.3. Pause the video as needed as you fill in all blanks. Circle your answer to each numbered exercise. Then press Play to continue listening to the video.

Objective A Add decimals.

Write the steps used to add or subtract decimals.

Adding or Subtracting Decimals
Step 1:
Step 2:
Step 3:

Add.

Work with me.

1. $1.3 + 2.2$

Pause and work.

2. $24.6 + 2.39 + 0.0678$

Play and check.

Section 4.3 Adding and Subtracting Decimals

Objective B Subtract decimals.

Subtract.

🔵 **Work with me.**

3. $18 - 2.7$

🔵 **Work with me.**

4. Subtract 6.7 from 23.

Objective C Estimate when adding or subtracting decimals.

Subtract. Do a quick check by estimating.

🔵 **Work with me.**

5. 1000
   -123.4

   Check:

⓫ Check subtraction by _____.

Objective D Solve problems that involve adding or subtracting decimals.

Solve the application.

🔵 **Work with me.**

6. A landscape architect is planning a border for a flower garden shaped like a triangle. The sides of the garden measure 12.4 feet, 29.34 feet, and 25.7 feet. Find the amount of border material needed.

Section 4.4 Multiplying Decimals and Circumference of a Circle

Complete the outline as you view Lecture Video 4.4. Pause the video as needed as you fill in all blanks. Circle your answer to each numbered exercise. Then press Play to continue listening to the video.

Objective A Multiply decimals.

Write the steps used to multiply decimals.

Multiplying Decimals

Step 1:

Step 2:

Multiply.

Work with me.

1.   1.2
    $\times 0.5$

Objective B Estimate when multiplying decimals.

Estimate the product to double check the actual product.

Work with me.

2.   1.0047
    $\times\ \ \ 8.2$

Section 4.4 Multiplying Decimals and Circumference of a Circle

Objective C Multiply by powers of 10.

Multiplying Decimals by Powers of 10 such as 10, 100, 1000, 10,000 …
Move the decimal point to the right the same number of places as there are zeros in the power of 10.

Multiply.

🔘 **Work with me.**

3. 7.093×100

Multiplying Decimals by Powers of 10 such as 0.1, 0.01, 0.001, 0.0001, …
Move the decimal point to the left the same number of places as there are decimal places in the power of 10.

Multiply.

🔘 **Work with me.**

4. 37.62×0.001

Objective D Find the circumference of a circle.

Circumference of a Circle

Circumference = $2 \cdot \pi \cdot$ radius

or

Circumference = $\pi \cdot$ diameter

$\pi \approx 3.14$ or $\pi \approx \dfrac{22}{7}$

Section 4.4 Multiplying Decimals and Circumference of a Circle

> ① The _____ of a circle is the distance around the circle.

Work with me.

5. Find the actual and approximate circumference of the circle.

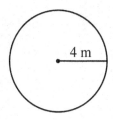

Objective E Solve problems by multiplying decimals.

Solve the application.

Work with me.

6. A 1-ounce serving of cream cheese contains 6.2 grams of saturated fat. How much saturated fat is in 4 ounces of cream cheese?

Basic College Mathematics Sixth Edition, Elayn Martin-Gay   Sec. 4.5

Section 4.5 Dividing Decimals and Order of Operations

Complete the outline as you view Lecture Video 4.5. Pause the video as needed as you fill in all blanks. Circle your answer to each numbered exercise. Then press Play to continue listening to the video.

Objective A Divide decimals.

Write the steps used to divide by a whole number.

Dividing by a Whole Number
Step 1:
Step 2:

Divide.

Work with me.

1. $5\overline{)0.47}$

Write the steps used to divide by a decimal.

Dividing by a Decimal
Step 1:
Step 2:
Step 3:

Divide and check.

Pause and work.

2. $0.82\overline{)4.756}$

Play and check.

Section 4.5 Dividing Decimals and Order of Operations

Divide. Round to the nearest hundredth.

Work with me.

3. $0.023\overline{)0.549}$

Objective B Estimate when dividing decimals.

Work with me.

Estimate the quotient to determine whether the actual answer is close to the estimate.

4. $5.5\overline{)36.3}$

Objective C Divide decimals by powers of 10.

Dividing Decimals by Powers of 10 such as 10, 100, or 1000
Move the decimal point in the dividend to the left the same number of places as there are zeros in the power of 10.

Divide.

Work with me.

5. $\dfrac{54.982}{100}$

Pause and work.

6. $\dfrac{12.9}{1000}$

Play and check.

Basic College Mathematics Sixth Edition, Elayn Martin-Gay   Sec. 4.5

Section 4.5 Dividing Decimals and Order of Operations

Objective D Solve problems by dividing decimals.

Solve the application.

▶ **Work with me.**

7. In the United States, an average child will wear down 730 crayons by his or her tenth birthday. Find the number of boxes of 64 crayons this is equivalent to. Round to the nearest tenth.

Objective E Review order of operations by simplifying expressions containing decimals.

⏸ Review the order of operations by writing the steps used below.

Order of Operations

Step 1:

Step 2:

Step 3:

Step 4:

▶ **Work with me.**

8. $0.7(6-2.5)$

⏸ **Pause and work.**

9. $\dfrac{7.8+1.1\times 100-3.6}{0.2}$

▶ **Play and check.**

64

Section 4.6 Fractions and Decimals

Complete the outline as you view Lecture Video 4.6. Pause ⏸ the video as needed as you fill in all blanks. Circle your answer to each numbered exercise. Then press Play ▶ to continue listening to the video.

Objective A Write fractions as decimals.

Writing Fractions as Decimals
To write a fraction as a decimal, divide the numerator by the denominator.

Write each fraction as a decimal.

● Work with me.

1. $\dfrac{3}{4}$

● Work with me.

2. $\dfrac{11}{12}$

Objective B Compare fractions and decimals.

● Work with me.

3. Compare: 1.38 $\dfrac{18}{13}$

Objective C Solve area problems containing fractions and decimals.

Area is measured in square units.

● Work with me.

4. Find the area of the rectangle.

0.62 yd

$\dfrac{2}{5}$ yd

Section 5.1 Ratios

Complete the outline as you view Lecture Video 5.1. Pause the video as needed as you fill in all blanks. Circle your answer to each numbered exercise. Then press Play to continue listening to the video.

Objective A Write ratios as fractions.

 A _____ is the quotient of two quantities.

Writing a Ratio as a Fraction
The order of the quantities is important when writing ratios. To write a ratio as a fraction, write the first number of the ratio as the numerator of the fraction and the second number as the denominator.

Write the ratio in fractional notation.

Work with me.

1. 23 to 10

Objective B Write ratios in simplest form.

Write the ratio in fractional notation in simplest form.

Work with me.

2. 16 to 24

Work with me.

3. 7.7 to 10

Work with me.

4. $3\frac{1}{2}$ to $12\frac{1}{4}$

Basic College Mathematics Sixth Edition, Elayn Martin-Gay   Sec. 5.1

Section 5.1 Ratios

⏸ **Pause and work.**

5. 24 days to 14 days

▶ Play and check.

Find the ratio of women to men in simplest form.

⏸ **Pause and work.**

6. 125 women and 100 men

▶ Play and check.

Basic College Mathematics Sixth Edition, Elayn Martin-Gay   Sec. 5.2

Section 5.2 Rates

Complete the outline as you view Lecture Video 5.2. Pause ⏸ the video as needed as you fill in all blanks. Circle your answer to each numbered exercise. Then press Play ▶ to continue listening to the video.

Objective A Write rates as fractions.

> ⏸ A special type of ratio is _____. _____ are used to compare different types of quantities.

Write the rates as fractions.

▶ **Work with me.**

1. 6 laser printers for 28 computers

⏸ **Pause and work.**

2. 5 shrubs every 15 feet

▶ **Play and check.**

Objective B Find unit rates.

> **Writing a Rate as a Unit Rate**
> To write a rate as a unit rate, divide the numerator of the rate by the denominator.

> ⏸ A _____ _____ is a rate with a denominator of 1.

Find the unit rate.

▶ **Work with me.**

3. 375 riders in 5 subway cars

⏸ **Pause and work.**

4. $1,000,000 lottery winnings paid over 20 years

▶ **Play and check.**

68

Copyright © 2019 Pearson Education, Inc.

Section 5.2 Rates

Objective C Find unit prices.

When a unit rate is "money per item," it is also called a unit price.

$$\text{unit price} = \frac{\text{price}}{\text{number of units}}$$

Find the best buy.

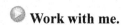 **Work with me.**

5. Soy sauce
   12 oz for $2.29
   8 oz for $1.49

Basic College Mathematics Sixth Edition, Elayn Martin-Gay   Sec. 5.3

Section 5.3 Proportions

Complete the outline as you view Lecture Video 5.3. Pause ⏸ the video as needed as you fill in all blanks. Circle your answer to each numbered exercise. Then press Play ▶ to continue listening to the video.

Objective A Write sentences as proportions.

⏸ A _____ is a statement that two ratios are equal.

⏸ The proportion $\frac{10}{12} = \frac{15}{18}$ is read "10 is to 12 _____ 15 is to 18."

Write sentence as a proportion.

▶ **Work with me.**

1. 10 diamonds is to 6 opals as 5 diamonds is to 3 opals.

Objective B Determine whether proportions are true.

Using Cross Products to Determine Whether Proportions Are True or False
$$\frac{a}{b} = \frac{c}{d}$$
Cross Products: $a \cdot d$, $b \cdot c$
If cross products are equal, then the proportion is true. If the cross products are not equal, the proportion is false.

Determine whether the proportions are true or false.

▶ **Work with me.**

2. $\frac{8}{6} = \frac{9}{7}$

⏸ **Pause and work.**

3. $\frac{9}{36} = \frac{2}{8}$

▶ **Play and check.**

Section 5.3 Proportions

Objective C Find an unknown number in a proportion.

Write the steps used to find an unknown value *n* in a proportion.

Finding an Unknown Value *n* in a Proportion

Step 1:

Step 2:

Find the value of the unknown number.

Work with me.

4. $\dfrac{n}{8} = \dfrac{50}{100}$

Pause and work.

5. $\dfrac{8}{\frac{1}{3}} = \dfrac{24}{n}$

Play and check.

Pause and work.

6. $\dfrac{n}{1\frac{1}{5}} = \dfrac{4\frac{1}{6}}{6\frac{2}{3}}$

Play and check.

Basic College Mathematics Sixth Edition, Elayn Martin-Gay   Sec. 5.4

Section 5.4 Proportions and Problem Solving

Complete the outline as you view Lecture Video 5.4. Pause the video as needed as you fill in all blanks. Circle your answer to each numbered exercise. Then press Play to continue listening to the video.

Objective A Solve problems by writing proportions.

Solve.

Work with me.

1. The school's computer lab goes through 5 reams of printer paper every 3 weeks. Find how long a case of printer paper is likely to last (a case of paper holds 8 reams of paper). Round to the nearest week.

Step 1: Understand

Step 2: Translate

Step 3: Solve

Step 4: Interpret

Pause and work.

2. There are 72 milligrams of cholesterol in a 3.5 ounce serving of lobster. How much cholesterol is in 5 ounces of lobster? Round to the nearest tenth of a milligram.

Play and check.
Pause and work.

3. Medication is prescribed in 7 out of every 10 hospital emergency room visits that involve an injury. If a large urban hospital had 620 emergency room visits involving an injury in the past month, how many of these visits would you expect included a prescription for medication?

Play and check.

Section 6.1 Introduction to Percent

Complete the outline as you view Lecture Video 6.1. Pause the video as needed as you fill in all blanks. Circle your answer to each numbered exercise. Then press Play to continue listening to the video.

Objective A Understand percent.

Percent means per one hundred. The "%" symbol is used to denote percent.

Work with me.

1. In a survey of 100 college students, 96 use the internet. What percent use the internet?

_____ means _____ hundred.

Objective B Write percents as decimals.

Writing a Percent as a Decimal
Replace the percent symbol with its decimal equivalent, 0.01; then multiply.
   43% = 43(0.01) = 0.43

Write the percent as an equivalent decimal.

Work with me.

2. 41%

Work with me.

3. 6%

Percent ⟶ Decimal
Drop the % symbol and move the decimal point 2 places to the left.

Work with me.

4. 100%

100% = 1

Basic College Mathematics Sixth Edition, Elayn Martin-Gay   Sec. 6.1

Section 6.1 Introduction to Percent

● **Work with me.**

5. 2.8%

● **Work with me.**

6. People take aspirin for a variety of reasons. The most common use of aspirin is to prevent heart disease, accounting for 38% of all aspirin use.

Objective C Write decimals as percents.

Writing a Decimal as a Percent
Multiply by 1 in the form of 100%.
$0.27 = 0.27(100\%) = 27\%$

Write the decimal as an equivalent percent.

● **Work with me.**

7. 0.22

Decimal ⟶ Percent
Move the decimal point 2 places to the right and attach a %.

● **Work with me.**

8. 0.056

● **Work with me.**

9. 3

⏸ **Pause and work.**

10. 0.7

● Play and check.

Basic College Mathematics Sixth Edition, Elayn Martin-Gay   Sec. 6.2

Section 6.2 Percents and Fractions

Complete the outline as you view Lecture Video 6.2. Pause the video as needed as you fill in all blanks. Circle your answer to each numbered exercise. Then press Play to continue listening to the video.

Objective A Write percents as fractions.

Writing a Percent as a Fraction
Replace the percent symbol with its fraction equivalent, $\frac{1}{100}$; then multiply. Don't forget to simplify the fraction if possible.

Write the percent as a fraction.

Work with me.

1. 4%

Work with me.

2. 175%

Work with me.

3. $10\frac{1}{3}\%$

Objective B Write fractions as percents.

Writing a Fraction as a Percent
Multiply by 1 in the form of 100%.

Write the fraction as a percent.

Work with me.

4. $\frac{7}{10}$

Section 6.2 Percents and Fractions

⏸ **Pause and work.**

5. $\dfrac{3}{8}$

▶ Play and check.

▶ **Work with me.**

6. $\dfrac{4}{15}$

Objective C Convert percents, decimals, and fractions.

Write a percent as a decimal:
$p\% = p(0.01)$

Write a percent as a fraction:
$p\% = p \cdot \dfrac{1}{100}$

Write a number as a percent:
$p = p \cdot 100\%$

Write the percent as an equivalent fraction and decimal.

▶ **Work with me.**

7. Africa: 14.2%

Basic College Mathematics Sixth Edition, Elayn Martin-Gay   Sec. 6.3

Section 6.3 Solving Percent Problems Using Equations

Complete the outline as you view Lecture Video 6.3. Pause the video as needed as you fill in all blanks. Circle your answer to each numbered exercise. Then press Play to continue listening to the video.

Objective A Write percents problems as equations.

> An _____ is a statement that uses an equal sign (=).

> of means multiplication (·)
> is means equals (=)
> what (or some equivalent) means the unknown number

Translate the statement to an equation.

Work with me.

1. 18% of 81 is what number?

Work with me.

2. What percent of 80 is 3.8?

Pause and work.

3. 1.2 is 12% of what number?

Play and check.

Objective B Solve percent problems.

> **Percent Equation**
> percent · base = amount

Section 6.3 Solving Percent Problems Using Equations

Find the value of n.

● **Work with me.**

4. $1.2 = 12\% \cdot n$

Translate the statement and then solve for n.

● **Work with me.**

5. 10% of 35 is what number?

❚❚ **Pause and work.**

6. 2.58 is what percent of 50?

● **Play and check.**

Basic College Mathematics Sixth Edition, Elayn Martin-Gay   Sec. 6.4

Section 6.4 Solving Percent Problems Using Proportions

Complete the outline as you view Lecture Video 6.4. Pause ⏸ the video as needed as you fill in all blanks. Circle your answer to each numbered exercise. Then press Play ▶ to continue listening to the video.

Objective A Write percents problems as proportions.

Percent Proportion

$$\frac{\text{amount}}{\text{base}} = \frac{\text{percent}}{100} \leftarrow \text{always 100}$$

or

$$\text{amount} \rightarrow \frac{a}{b} = \frac{p}{100} \leftarrow \text{percent}$$
$$\text{base} \rightarrow$$

Part of Proportion	How it's identified
Percent	% or percent
Base	Appears after "of"
Amount	Part compared to whole

Translate the statement into a proportion.

▶ **Work with me.**

1. 98% of 45 is what number?

⏸ The _____ usually appears after the word "of."

⏸ **Pause and work.**

2. What percent of 400 is 70?

▶ Play and check.

▶ **Work with me.**

3. 7.8 is 78% of what number?

79

Basic College Mathematics Sixth Edition, Elayn Martin-Gay   Sec. 6.4

Section 6.4 Solving Percent Problems Using Proportions

Objective B Solve percent problems.

Solve the proportion.

🔵 **Work with me.**

4. $\dfrac{7.8}{b} = \dfrac{78}{100}$

Translate the statement into a proportion and then solve.

🔵 **Work with me.**

5. What percent of 6 is 2.7?

⏸ **Pause and work.**

6. 20% of 48 is what number?

🔵 **Play and check.**

Basic College Mathematics Sixth Edition, Elayn Martin-Gay   Sec. 6.5

Section 6.5 Applications of Percent

Complete the outline as you view Lecture Video 6.5. Pause ⏸ the video as needed as you fill in all blanks. Circle your answer to each numbered exercise. Then press Play ▶ to continue listening to the video.

Objective A Solve applications involving percent.

Solve the percent problem.

▶ **Work with me.**

1. A family paid $26,250 as a down payment for a home. If this represents 15% of the price of the home, find the price of the home.

is means equals

of means multiplication

percent, p: look for the % symbol

base, b: usually follows the word "of"

amount, a: part compared to the whole

Objective B Find percent increase and percent decrease.

Percent of Increase

percent of increase = $\dfrac{\text{amount of increase}}{\text{original amount}}$

Then write the quotient as a percent.

Percent of Decrease

percent of decrease = $\dfrac{\text{amount of decrease}}{\text{original amount}}$

Then write the quotient as a percent.

Section 6.5 Applications of Percent

Determine the increase or decrease.

Work with me.

2. There are 150 calories in a cup of whole milk and only 84 in a cup of skim milk. In switching to skim milk, find the percent decrease in number of calories per cup.

Pause and work.

3. In 1940, the average size of a privately owned farm in the United States was 174 acres. In recent years, the average size of a privately owned farm in the United States has increased to 421 acres. Find the percent of increase.

Play and check.

Basic College Mathematics Sixth Edition, Elayn Martin-Gay   Sec. 6.6

Section 6.6 Percent and Problem Solving: Sales Tax, Commission, and Discount

Complete the outline as you view Lecture Video 6.6. Pause the video as needed as you fill in all blanks. Circle your answer to each numbered exercise. Then press Play to continue listening to the video.

Objective A Calculate sales tax and total price.

Sales Tax and Total Price
sales tax = tax rate · purchase price
total price = purchase price + sales tax

Calculate the purchase price.

Work with me.

1. The sales tax on the purchase of a futon is $24.25. If the tax rate is 5%, find the purchase price of the futon.

Objective B Calculate commissions.

Commission
commission = commission rate · sales

Determine the commission rate.

Work with me.

2. A salesperson earned a commission of $1380.40 for selling $9860 worth of paper products. Find the commission rate.

Objective C Calculate discount and sale price.

Discount and Sale Price
amount of discount = discount rate · original price
sale price = original price − amount of discount

Section 6.6 Percent and Problem Solving: Sales Tax, Commission, and Discount

Find the amount of the discount and the sale price.

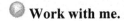 **Work with me.**

3. A $300 fax machine is on sale for 15% off. Find the amount of discount and the sale price.

Section 6.7 Percent and Problem Solving: Interest

Complete the outline as you view Lecture Video 6.7. Pause the video as needed as you fill in all blanks. Circle your answer to each numbered exercise. Then press Play to continue listening to the video.

Objective A Calculate simple interest.

Simple Interest
Simple Interest = Principal · Rate · time
$$I = P \cdot R \cdot T$$
where the rate is understood to be per year and time is in years.

Finding the Total Amount of a Loan or Investment
total amount (paid or received) = principal + interest

Determine the interest paid on the loan and the total amount paid back.

Work with me.

1. A company borrows $162,500 for 5 years at a simple interest of 12.5%. Find the interest paid on the loan and the total amount paid back.

Objective B Calculate compound interest.

Compound Interest Formula
The total amount A in an account is given by
$$A = P\left(1 + \frac{r}{n}\right)^{n \cdot t}$$
where P is the principal, r is the interest rate written as a decimal, t is the length of time in years, and n is the number of times compounded per year.

Find the total amount in the account.

Work with me.

2. Find the total amount in this account. $6150 is compounded semiannually at a rate of 14% for 15 years.

Basic College Mathematics Sixth Edition, Elayn Martin-Gay   Sec. 6.7

Section 6.7 Percent and Problem Solving: Interest

Objective C Calculate monthly payments.

Finding the Monthly Payment of a Loan

$$\text{monthly payment} = \frac{\text{principal} + \text{interest}}{\text{total number of payments}}$$

Determine the monthly payment.

Work with me.

3. $20,000 is borrowed for 4 years. If the interest on the loan is $10,588.40, find the monthly payment.

Section 7.1 Length: U.S. and Metric Systems of Measurement

Complete the outline as you view Lecture Video 7.1. Pause the video as needed as you fill in all blanks. Circle your answer to each numbered exercise. Then press Play to continue listening to the video.

Objective A Define U.S. units of length and convert from one unit to another.

U.S. Units of Length
12 inches (in.) = 1 foot (ft)
3 feet = 1 yard (yd)
36 inches = 1 yard
5280 feet = 1 mile (mi)

A _____ _____ is a fraction that equals 1.

Perform the conversion as indicated.

Work with me.

1. 60 in. to feet

Pause and work.

2. $8\frac{1}{2}$ ft to inches

Play and check.

Work with me.

3. 10 ft to yards

Objective B Use mixed units of length.

Perform the conversion as indicated.

Work with me.

4. 5 ft 2 in. to inches

Section 7.1 Length: U.S. and Metric Systems of Measurement

Objective C Perform arithmetic operations on U.S. units of length.

Perform the indicated operation.

 Work with me.

5.   12 yd 2 ft
   + 9 yd 2 ft
   ―――――――

Objective D Define metric units of length and convert from one unit to another.

The basic unit of length in the metric system is the _____.

Metric Unit of Length
1 kilometer (km) = 1000 meters (m)
1 hectometer (hm) = 100 m
1 dekameter (dam) = 10 m
1 meter (m) = 1 m
1 decimeter (dm) = 1/10 m or 0.1 m
1 centimeter (cm) = 1/100 m or 0.01 m
1 millimeter (mm) = 1/1000 m or 0.001 m

Perform the conversion as indicated.

 Work with me.

6. 1500 cm to meters

 Pause and work.

7. 0.04 m to millimeters

 Play and check.

Section 7.1 Length: U.S. and Metric Systems of Measurement

Objective E **Perform arithmetic operations on metric units of length.**

Perform the indicated operation.

Work with me.

8. 24.8 mm − 1.19 cm

Basic College Mathematics Sixth Edition, Elayn Martin-Gay   Sec. 7.2

Section 7.2 Weight and Mass: U.S. and Metric Systems of Measurement

Complete the outline as you view Lecture Video 7.2. Pause the video as needed as you fill in all blanks. Circle your answer to each numbered exercise. Then press Play to continue listening to the video.

Objective A Define U.S. units of weight and convert from one unit to another.

U.S. Units of Weight
16 ounces (oz) = 1 pound (lb)
2000 pounds = 1 ton

A unit fraction is a fraction that equals 1.

Perform the conversion as indicated.

Work with me.

1. 60 ounces to pounds

Pause and work.

2. 4.9 tons to pounds

Play and check.

Work with me.

3. 89 oz to lb and oz

Objective B Perform arithmetic operations on units of weight.

Perform the indicated operation.

Work with me.

4. 12 lb 4 oz
   − 3 lb 9 oz
   ─────────

Section 7.2 Weight and Mass: U.S. and Metric Systems of Measurement

Objective C Define metric units of mass and convert from one unit to another.

> The basic unit of mass in the metric system is the _____.

Metric Unit of Mass
1 kilogram (kg) = 1000 grams (g)
1 hectogram (hg) = 100 g
1 dekagram (dag) = 10 g
1 gram (g) = 1 g
1 decigram (dg) = 1/10 g or 0.1 g
1 centigram (cg) = 1/100 g or 0.01 g
1 milligram (mg) = 1/1000g or 0.001g

Perform the conversion as indicated.

Work with me.

5. 4 g to milligrams

Pause and work.

6. 6.3 g to kilograms

Play and check.

Objective D Perform arithmetic operations on units of mass.

Perform the indicated operation.

Work with me.

7. 9 g − 7150 mg

Section 7.3 Capacity: U.S. and Metric Systems of Measurement

Complete the outline as you view Lecture Video 7.3. Pause the video as needed as you fill in all blanks. Circle your answer to each numbered exercise. Then press Play to continue listening to the video.

Objective A Define U.S. units of capacity and convert from one unit to another.

U.S. Units of Capacity
8 fluid ounces (fl oz) = 1 cup (c)
2 cups = 1 pint (pt)
2 pints = 1 quart (qt)
4 quarts = 1 gallon (gal)

A unit fraction is a fraction that equals 1.

Perform the conversion as indicated.

Work with me.

1. 14 qt to gallons

Pause and work.

2. 42 cups to quarts

Play and check.

Work with me.

3. 58 qt to gal and qt

Objective B Perform arithmetic operations on U.S. units of capacity.

Perform the indicated operation.

Work with me.

4.   3 gal
   −1 gal 3 qt
   ―――――――

Basic College Mathematics Sixth Edition, Elayn Martin-Gay   Sec. 7.3

Section 7.3 Capacity: U.S. and Metric Systems of Measurement

Objective C Define metric units of capacity and convert from one unit to another.

The basic unit of capacity in the metric system is the _____.

Metric Unit of Capacity
1 kiloliter (kl) = 1000 liters (L)
1 hectoliter (hl) = 100 L
1 dekaliter (dal) = 10 L
1 liter (L) = 1 L
1 deciliter (dl) = 1/10 L or 0.1 L
1 centiliter (cl) = 1/100 L or 0.01 L
1 milliliter (ml) = 1/1000 L or 0.001 L

Perform the conversion as indicated.

Work with me.

5. 5600 ml to liters

Pause and work.

6. 0.16 kl to liters

Play and check.

Objective D Perform arithmetic operations on metric units of capacity.

Perform the indicated operation.

Work with me.

7. 2700 ml + 1.8 L

Basic College Mathematics Sixth Edition, Elayn Martin-Gay   Sec. 7.4

Section 7.4 Conversions Between the U.S. and Metric Systems

Complete the outline as you view Lecture Video 7.4. Pause ⏸ the video as needed as you fill in all blanks. Circle your answer to each numbered exercise. Then press Play ▶ to continue listening to the video.

Objective A Convert between the U.S. and metric systems.

Length:	Capacity:	Weight (mass):
metric U.S. System	metric U.S. System	metric U.S. System
$1\text{ m} \approx 1.09\text{ yd}$	$1\text{ L} \approx 1.06\text{ qt}$	$1\text{ kg} \approx 2.20\text{ qt}$
$1\text{ m} \approx 3.28\text{ ft}$	$1\text{ L} \approx 0.26\text{ gal}$	$1\text{ g} \approx 0.04\text{ oz}$
$1\text{ km} \approx 0.62\text{ mi}$	$3.79\text{ L} \approx 1\text{ gal}$	$0.45\text{ kg} \approx 1\text{ lb}$
$2.54\text{ cm} = 1\text{ in.}$	$0.95\text{ L} \approx 1\text{ qt}$	$28.35\text{ g} \approx 1\text{ oz}$
$0.30\text{ m} \approx 1\text{ ft}$	$29.57\text{ ml} \approx 1\text{ fl oz}$	
$1.61\text{ km} \approx 1\text{ mi}$		

A unit fraction is a fraction that equals 1.

Perform the conversion as indicated.

▶ **Work with me.**

1. 86 inches to cm

⏸ **Pause and work.**

2. 14.5 L to gallons

▶ **Play and check.**

▶ **Work with me.**

3. For an average adult, the weight of a right lung is greater than the weight of a left lung. If the right lung weighs 1.5 pounds and the left lung weighs 1.25 pounds, find the difference in grams.

Basic College Mathematics Sixth Edition, Elayn Martin-Gay   Sec. 7.5

Section 7.5 Temperature: U.S. and Metric Systems of Measurement

Complete the outline as you view Lecture Video 7.5. Pause the video as needed as you fill in all blanks. Circle your answer to each numbered exercise. Then press Play to continue listening to the video.

Objective A Convert temperatures from degrees Celsius to degrees Fahrenheit.

Converting Celsius to Fahrenheit
$F = \frac{9}{5}C + 32$ or $F = 1.8C + 32$

Work with me.

1. A weather forecaster in Caracas predicts a high temperature of 27°C. Find this measurement in degrees Fahrenheit.

Objective B Convert temperatures from degrees Fahrenheit to degrees Celsius.

Converting Fahrenheit to Celsius
$C = \frac{5}{9}(F - 32)$

Perform the indicated conversion.

Work with me.

2. 77°F to degrees Celsius

Water freezes at 32°F = 0°C
Water boils at 212°F = 100°C

Section 7.6 Energy: U.S. and Metric Systems of Measurement

Complete the outline as you view Lecture Video 7.6. Pause the video as needed as you fill in all blanks. Circle your answer to each numbered exercise. Then press Play to continue listening to the video.

Objective A Define and use U.S. units of energy and convert from one unit to another.

_____ can be measured in foot-pounds (ft-lb).

One foot-pound (ft-lb) is the amount of energy needed to lift a 1-pound object a distance of 1 foot.

Work with me.

1. How much energy is required to lift a 3-pound math textbook 380 feet up a hill?

_____ is a form of energy.

In the U.S. system of measurement, heat is measured in British Thermal Units (BTU). A BTU is the amount of heat required to raise the temperature of 1 pound of water 1 degree Fahrenheit.

Work with me.

2. Convert 30 BTU to foot-pounds.

Objective B Define and use metric units of energy.

A calorie (cal) is the amount of heat required to raise the temperature of 1 kilogram of water 1 degree Celsius.

Work with me.

3. Approximately 300 calories are burned each hour skipping rope. How many calories are required to skip rope $\frac{1}{2}$ of an hour each day for 5 days?

Basic College Mathematics Sixth Edition, Elayn Martin-Gay   Sec. 8.1

Section 8.1 Lines and Angles

Complete the outline as you view Lecture Video 8.1. Pause the video as needed as you fill in all blanks. Circle your answer to each numbered exercise. Then press Play to continue listening to the video.

Objective A **Identify lines, line segments, rays, and angles.**

A _____ extends indefinitely in all directions.

A _____ is a flat surface that extends indefinitely.

A _____ has no length, no width, and no height, but it does have location.

A _____ is a set of points extending indefinitely in two directions.

Work with me.

1. Identify the line.

A _____ is a part of a line with an end point.

Work with me.

2. Identify the ray.

Two rays with a common endpoint make up an _____.

The common endpoint of two rays is called the _____.

Objective B **Classify angles as acute, right, obtuse, or straight.**

An angle can be measured in _____.

An angle that measures 180° is called a _____ _____.

97

Copyright © 2019 Pearson Education, Inc.

Section 8.1 Lines and Angles

| An angle that measures 90° is called a _____ _____. |

| An _____ _____ measures between 0° and 90°. |

| An _____ _____ measures between 90° and 180°. |

Work with me.

3. Classify the angle as a straight angle, a right angle, an acute angle, or an obtuse angle.

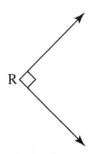

Pause and work.

4. Classify the angle as a straight angle, a right angle, an acute angle, or an obtuse angle.

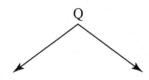

Play and check.

Objective C Identify complementary and supplementary angles.

| Two angles that have a sum of 90° are called complementary angles. We say that each angle is the complement of the other. |

| Two angles that have a sum of 180° are called supplementary angles. We say that each angle is the supplement of the other. |

Section 8.1 Lines and Angles

● **Work with me.**

5. Find the complement of a 23° angle.

● **Pause and work.**

6. Find the supplement of a 17° angle.

● **Play and check.**

Objective D Find measures of angles.

| _____ _____ are lines that never meet. |

| _____ _____ meet at a point. |

| If right angles are formed when two lines intersect, the two lines are _____ . |

| When two lines intersect, the angles opposite each other are called _____ _____ . |

| _____ _____ have the same measure. |

● **Work with me.**

7. Find the measure of the desired angles.

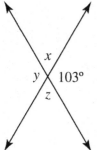

| _____ _____ share a common side. |

Section 8.1 Lines and Angles

Pause and work.

8. Find the measure of ∠ABC and ∠DBA.

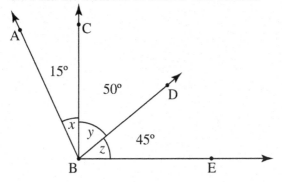

Play and check.

Parallel Lines Cut by a Transversal
If two parallel lines are cut by a transversal, then the measures of corresponding angles are equal and the measures of the alternate interior angles are equal.

Work with me.

9. Find the measure of the desired angles.

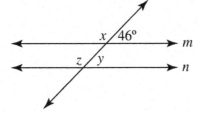

Section 8.2 Plane Figures and Solids

Complete the outline as you view Lecture Video 8.2. Pause the video as needed as you fill in all blanks. Circle your answer to each numbered exercise. Then press Play to continue listening to the video.

Objective A Identify plane figures.

A plane figure is a figure that lies on a plane. Plane figures, like planes, have length and width but no thickness or depth.

A polygon is a closed plane figure that basically consists of three or more line segments that meet at their end points.

The sum of the measures of the angles of a triangle is 180°.

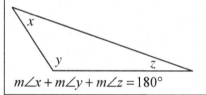

$m\angle x + m\angle y + m\angle z = 180°$

Work with me.

1. Find the measure of angle x.

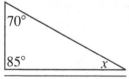

Pause and work.

2. Find the measure of angle x.

Play and check.

Basic College Mathematics Sixth Edition, Elayn Martin-Gay   Sec. 8.2

Section 8.2 Plane Figures and Solids

A right angle measures 90°.

A _____ is a polygon with 4 sides.

A parallelogram is a special quadrilateral with opposite sides parallel and equal in length.

A rectangle is a special parallelogram that has four right angles.

A square is a special rectangle that has all four sides equal in length.

A rhombus is a special parallelogram that has all four sides equal in length.

A trapezoid is a quadrilateral with exactly one pair of opposite sides parallel.

parallel sides

Section 8.2 Plane Figures and Solids

Work with me.

3. A quadrilateral with opposite sides parallel is a(n) ...

Pause and work.

4. True or False: A rectangle is also a parallelogram.

Play and check.

A circle is a plane figure that consists of all points that are the same fixed distance from a point c. The point c is called the center of the circle.

The radius of a circle is the distance from the center of the circle to any point on the circle.

The diameter of a circle is the distance across the circle passing through the center. Notice that the diameter is twice the radius, and the radius is half the diameter.

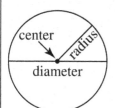

Work with me.

5. Find the diameter of the circle.

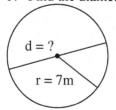

Objective B Identify solids.

A solid is a figure that lies in space. Solids have length, width, and height or depth.

Section 8.2 Plane Figures and Solids

A rectangular solid is a solid that consists of six sides, or faces, all of which are rectangles.

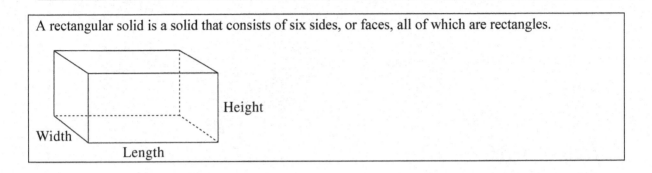

A cube is a rectangular solid whose six sides are squares.

A pyramid is shown below. The pyramids we will study have square bases and heights that are perpendicular to their base.

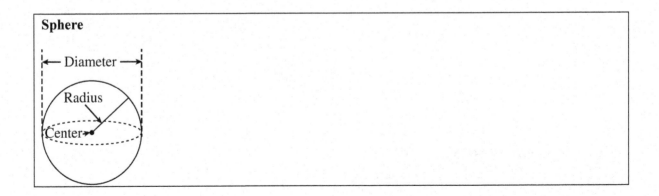

Sphere

Section 8.2 Plane Figures and Solids

The cylinders we will study have bases that are in the shape of circles and heights that are perpendicular to their base.

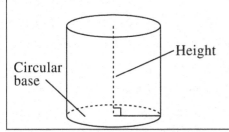

The cones we will study have bases that are circles and heights that are perpendicular to their base.

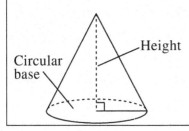

Work with me.

6. Identify the figure.

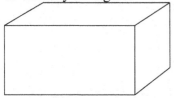

Pause and work.

7. Identify the figure.

Play and check.

Basic College Mathematics Sixth Edition, Elayn Martin-Gay   Sec. 8.3

Section 8.3 Perimeter

Complete the outline as you view Lecture Video 8.3. Pause ⏸ the video as needed as you fill in all blanks. Circle your answer to each numbered exercise. Then press Play ▶ to continue listening to the video.

Objective A Use formulas to find perimeter.

⏸ The perimeters of some special geometric figures form patterns, and these patterns are given as _____.

Perimeter of a Rectangle

Perimeter = 2 · length + 2 · width
In symbols, this can be written as $P = 2l + 2w$.

length

width   width

length

Perimeter of a Square

Perimeter = side + side + side + side
         = 4 · side
In symbols, $P = 4s$

side

side   side

side

Perimeter of a Triangle

Perimeter = side a + side b + side c
In symbols, $P = a + b + c$

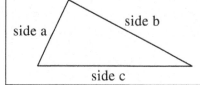

Section 8.3 Perimeter

🔵 **Work with me.**

1. Find the perimeter.

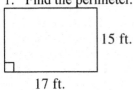

🔘 **Pause and work.**

2. Find the perimeter.

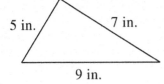

🔵 **Play and check.**

Perimeter of a Polygon
The perimeter of a polygon is the sum of the lengths of its sides.

🔘 **Pause and work.**

3. Find the perimeter.

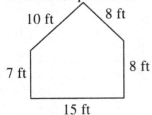

🔵 **Play and check.**

Basic College Mathematics Sixth Edition, Elayn Martin-Gay   Sec. 8.3

Section 8.3 Perimeter

Pause and work.

4. Find the perimeter.

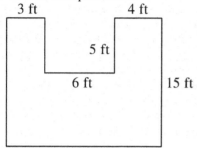

Play and check.

Pause and work.

5. Find the perimeter of the top of a square compact disc case if the length of one side is 7 inches.

Play and check.

Objective B Use formulas to find circumferences.

Circumference of a Circle

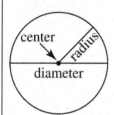

Circumference = $2 \cdot \pi \cdot$ radius or
Circumference = $\pi \cdot$ diameter

In symbols, $C = 2\pi r$ or $C = \pi d$ where $\pi \approx 3.14$ or $\pi \approx \dfrac{22}{7}$

Section 8.3 Perimeter

 Work with me.

6. Find the circumference.

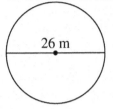

Basic College Mathematics Sixth Edition, Elayn Martin-Gay   Sec. 8.4

Section 8.4 Area

Complete the outline as you view Lecture Video 8.4. Pause ⏸ the video as needed as you fill in all blanks. Circle your answer to each numbered exercise. Then press Play ▶ to continue listening to the video.

Objective A Find the areas of geometric figures.

⏸ _____ measures the number of square units that cover a plane figure.

Rectangle

width
length

Area of rectangle: Area = length · width

$A = lw$

Square

side
side

Area of square: Area = side · side

$A = s \cdot s = s^2$

Triangle

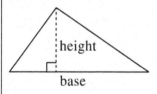
height
base

Area of triangle: Area = $\frac{1}{2}$ base · height

$A = \frac{1}{2}bh$

110

Copyright © 2019 Pearson Education, Inc.

Basic College Mathematics Sixth Edition, Elayn Martin-Gay   Sec. 8.4

Section 8.4 Area

Parallelogram

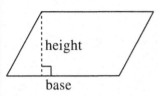

Area of parallelogram: Area = base · height

$A = bh$

Trapezoid

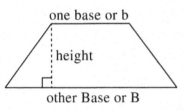

Area of trapezoid: Area = $\frac{1}{2}$(one base + other base) · height

$A = \frac{1}{2}(b + B)h$

⓫ _____ is measured in units. _____ is measured in square units.

▶ **Work with me.**

1. Find the area.

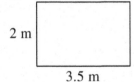

Section 8.4 Area

⏸ **Pause and work.**

2. Find the area.

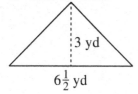

▶ Play and check.

⏸ **Pause and work.**

3. Find the area.

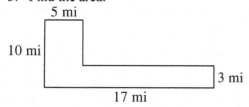

▶ Play and check.

Area Formula of a Circle

Area of a circle

Area = $\pi(\text{radius})^2$

$A = \pi r^2$

(A fraction approximation for π is $\frac{22}{7}$.)

(A decimal approximation for π is 3.14.)

Basic College Mathematics Sixth Edition, Elayn Martin-Gay   Sec. 8.4

Section 8.4 Area

___

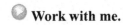 **Work with me.**

4. Find the area.

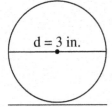

⚾ $\frac{1}{2}(\text{diameter}) = \text{radius}$

⚾ **Pause and work.**

5. A drapery panel measures 6 ft by 7 ft. Find how many square feet of material are needed for four panels.

⚾ Play and check.

113

Basic College Mathematics Sixth Edition, Elayn Martin-Gay   Sec. 8.5

Section 8.5 Volume

Complete the outline as you view Lecture Video 8.5. Pause ⏸ the video as needed as you fill in all blanks. Circle your answer to each numbered exercise. Then press Play ▶ to continue listening to the video.

Objective A Find the volume of solids.

Rectangular Solid

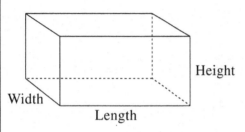

Volume = length · width · height
$V = l \cdot w \cdot h$

Cube

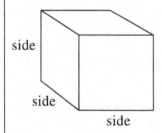

Volume = side · side · side
$V = s^3$

Sphere

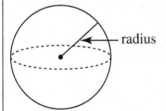

Volume = $\frac{4}{3} \cdot \pi \cdot (\text{radius})^3$
$V = \frac{4}{3} \cdot \pi \cdot r^3$

Section 8.5 Volume

Circular Cylinder

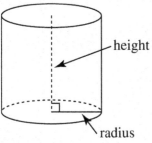

Volume = $\pi \cdot (\text{radius})^2 \cdot \text{height}$
$V = \pi \cdot r^2 \cdot h$

Cone

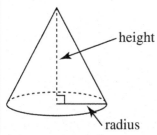

Volume = $\frac{1}{3} \cdot \pi \cdot (\text{radius})^2 \cdot \text{height}$
$V = \frac{1}{3} \cdot \pi \cdot r^2 \cdot h$

Square-Based Pyramid

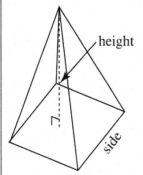

Volume = $\frac{1}{3} \cdot (\text{side})^2 \cdot \text{height}$
$V = \frac{1}{3} \cdot \pi \cdot s^2 \cdot h$

Section 8.5 Volume

▶ **Work with me.**

1. Find the volume of the solid.

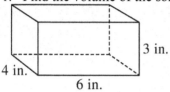

▮▮ **Pause and work.**

2. Find the volume of the solid.

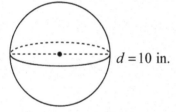

▶ Play and check.

▮▮ **Pause and work.**

3. Find the volume of the solid.

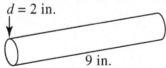

▶ Play and check.

Section 8.5 Volume

Pause and work.

4. Find the volume of the solid.

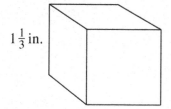

$1\frac{1}{3}$ in.

Play and check.

Pause and work.

5. Find the exact volume of a waffle ice cream cone with a 3-in. diameter and a height of 7 inches.

Play and check.

Basic College Mathematics Sixth Edition, Elayn Martin-Gay   Sec. 8.6

Section 8.6 Square Roots and the Pythagorean Theorem

Complete the outline as you view Lecture Video 8.6. Pause ⏸ the video as needed as you fill in all blanks. Circle your answer to each numbered exercise. Then press Play ▶ to continue listening to the video.

Objective A Find the square root of a number.

⏸ The reverse process of squaring is finding a _____.

We use the notation $\sqrt{}$, called a radical sign, to indicate the positive square root of a nonnegative number.

Square Root of a Number.

The square root, $\sqrt{}$, of a positive number a is the positive number b whose square is a. In symbols,
$\sqrt{a} = b$, if $b^2 = a$.
Also, $\sqrt{0} = 0$.

Find the square root.

▶ **Work with me.**

1. $\sqrt{4}$

⏸ **Pause and work.**

2. $\sqrt{121}$

▶ **Play and check.**

⏸ **Pause and work.**

3. $\sqrt{\dfrac{1}{81}}$

▶ **Play and check.**

Section 8.6 Square Roots and the Pythagorean Theorem

Objective B Approximate square roots.

🔵 **Work with me.**

4. $\sqrt{38}$ is between what two whole numbers?

⏸ **Pause and work.**

5. $\sqrt{15}$ is between what two whole numbers? Also, find the approximation for $\sqrt{15}$ to three decimal places.

🔵 Play and check.

Objective C Use the Pythagorean Theorem.

Pythagorean Theorem
In any right triangle,
$$(\text{leg})^2 + (\text{other leg})^2 = (\text{hypotenuse})^2$$

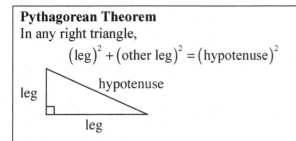

Finding an Unknown Length of a Right Triangle
$$\text{hypotenuse} = \sqrt{(\text{leg})^2 + (\text{other leg})^2}$$
or
$$\text{leg} = \sqrt{(\text{hypotenuse})^2 - (\text{other leg})^2}$$

⏸ The Pythagorean Theorem only applies to _____ _____.

Section 8.6 Square Roots and the Pythagorean Theorem

▶ **Work with me.**

6. Find the third length of the triangle.

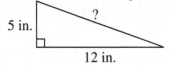

▐▐ **Pause and work.**

7. Find the third length of the triangle.
   hypotenuse = 2
   leg = 1

▶ **Play and check.**

Section 8.7 Congruent and Similar Triangles

Complete the outline as you view Lecture Video 8.7. Pause the video as needed as you fill in all blanks. Circle your answer to each numbered exercise. Then press Play to continue listening to the video.

Objective A Decide whether two triangles are congruent.

_____ _____ have the same shape and same size.

Congruent triangles: Corresponding angles have the same measure.
Corresponding sides have the same length.

Angle-Side-Angle (ASA)
If the measures of two angles of a triangle equal the measures of two angles of another triangle, and the lengths of the sides between each pair of angles are equal, the triangles are congruent.

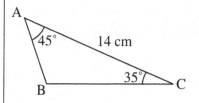 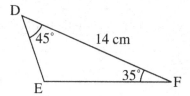

Side-Side-Side (SSS)
If the lengths of the three sides of a triangle equal the lengths of the corresponding sides of another triangle, the triangles are congruent.

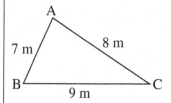 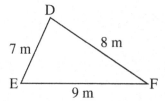

Side-Angle-Side (SAS)
If the lengths of two sides of a triangle equal the lengths of corresponding sides of another triangle, and the measures of the angles between each pair of sides are equal, the triangles are congruent.

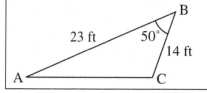 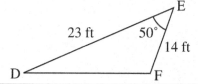

Basic College Mathematics Sixth Edition, Elayn Martin-Gay   Sec. 8.7

Section 8.7 Congruent and Similar Triangles

Work with me.

1. Determine whether the two triangles are congruent.

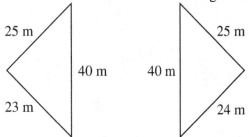

Objective B Find the ratio of corresponding sides in similar triangles.

_____ _____ have the same shape but not necessarily the same size.

Similar triangles: Corresponding angles have the same measure.
Corresponding sides are in proportion.

Work with me.

2. Find the ratios of the corresponding sides of the similar triangles

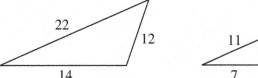

Objective C Find unknown lengths of sides in similar triangles.

Work with me.

3. Find the length of side *n*.

 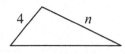

122

Section 8.7 Congruent and Similar Triangles

Pause and work.

4. If a 30-foot tree casts an 18-foot shadow, find the length of the shadow cast by a 24-foot tree.

Play and check.

Basic College Mathematics Sixth Edition, Elayn Martin-Gay   Sec. 9.1

Section 9.1 Reading Pictographs, Bar Graphs, Histograms, and Line Graphs

Complete the outline as you view Lecture Video 9.1. Pause the video as needed as you fill in all blanks. Circle your answer to each numbered exercise. Then press Play to continue listening to the video.

Objective A Read pictographs.

> A _____ is a graph that uses pictures or symbols. The _____ tells you what a symbol stands for.

Use the pictograph to answer the questions.

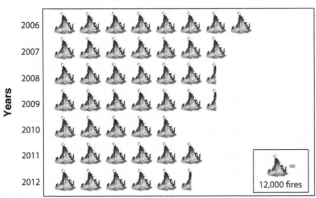

Work with me.

1. Approximate the number of wildfires in 2012.

Pause and work.

2. Which year, of the years shown, had the most wildfires?

Play and check.

Pause and work.

3. What was the amount of increase in wildfires from 2010 to 2011.

Play and check.

124

Copyright © 2019 Pearson Education, Inc.

Basic College Mathematics Sixth Edition, Elayn Martin-Gay   Sec. 9.1

Section 9.1 Reading Pictographs, Bar Graphs, Histograms, and Line Graphs

⏸ **Pause and work.**

4. What was the average annual number of wildfires from 2010 to 2012?

▶ **Play and check.**

Objective B Read and construct bar graphs.

⏸ _____ _____ have horizontal or vertical bars to show data.

Use the data to answer the question.

Fiber Content of Selected Foods	
Food	**Grams of Total Fiber**
Kidney beans $\left(\frac{1}{2}c\right)$	4.5
Oatmeal, cooked $\left(\frac{3}{4}c\right)$	3.0
Peanut butter, chunky (2 tbsp)	1.5
Popcorn (1 c)	1.0
Potato, baked with skin (1 med)	4.0
Whole wheat bread (1 slice)	2.5

▶ **Work with me.**

5. Draw a bar graph.

Basic College Mathematics Sixth Edition, Elayn Martin-Gay   Sec. 9.1

Section 9.1 Reading Pictographs, Bar Graphs, Histograms, and Line Graphs

Objective C **Read and construct histograms.**

A _____ is a special bar graph.

A histogram is different from a bar graph in that the width of each bar stands for a range of numbers, called a _____ _____.

The height of each bar is how many times a number occurs in the class and is called the _____ _____.

Use the histogram to answer the questions.

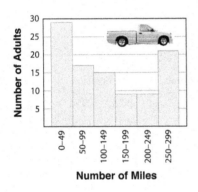

Work with me.

6. How many adults drive fewer than 150 miles per week.

Pause and work.

7. How many more adults drive 250–299 miles per week than 200–249 miles per week?

Play and check.

126

Copyright © 2019 Pearson Education, Inc.

Section 9.1 Reading Pictographs, Bar Graphs, Histograms, and Line Graphs

Use the data to answer the following questions

78	84	91	93	97
97	95	85	95	96
101	89	92	89	100

Work with me.

8. Create a frequency table.

Work with me.

9. Create a histogram.

Objective D Read line graphs.

Use the line graph to answer the following questions

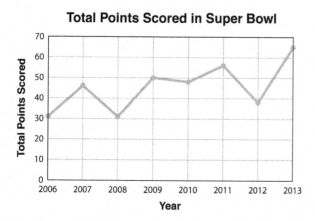

Total Points Scored in Super Bowl

Basic College Mathematics Sixth Edition, Elayn Martin-Gay   Sec. 9.1

Section 9.1 Reading Pictographs, Bar Graphs, Histograms, and Line Graphs

🔴 **Work with me.**

10. During which year(s) shown was the total score in the Super Bowl the lowest?

⏸ **Pause and work.**

11. During which year(s) was the total score in the Super Bowl less than 40?

🔴 Play and check.

⏸ **Pause and work.**

12. Find the total points scored in the Super Bowl in 2009?

🔴 Play and check.

⏸ **Pause and work.**

13. During which year(s) shown were the total points scored in the Super Bowl greater than 50?

🔴 Play and check.

Section 9.2 Reading Circle Graphs

Complete the outline as you view Lecture Video 9.2. Pause the video as needed as you fill in all blanks. Circle your answer to each numbered exercise. Then press Play to continue listening to the video.

Objective A Read circle graphs.

A survey of 700 college students was taken asking where they lived while attending college. Answer the following questions by reading the circle graph.

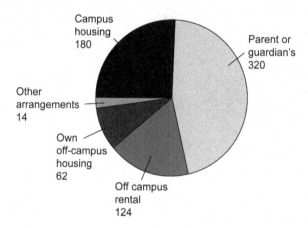

Work with me.

1. Where do most of these college students live?

Work with me.

2. Find the ratio of students living in campus housing to total students. Simplify your answer.

Basic College Mathematics Sixth Edition, Elayn Martin-Gay   Sec. 9.2

Section 9.2 Reading Circle Graphs

The circle graph shows the percent of types of books available at Midway Memorial Library. Answer the following questions by reading the circle graph.

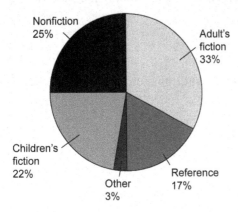

⚪ **Work with me.**

3. What percent of books are classified as some type of fiction?

⏸ **Pause and work.**

4. What is the second-largest category of books?

⚪ **Play and check.**

⚪ **Work with me.**

5. Suppose the library has 125,600 books, how many of these are nonfiction?

Basic College Mathematics Sixth Edition, Elayn Martin-Gay   Sec. 9.2

Section 9.2 Reading Circle Graphs

Objective B Draw circle graphs.

Work with me.

6. Find the degrees in each sector and draw the corresponding circle graph. Round to the nearest whole degree.

Types of Apples Grown in Washington State		
Type of Apple	**Percent**	**Degrees in Sector**
Red Delicious	37%	
Golden Delicious	13%	
Fuji	14%	
Gala	15%	
Granny Smith	12%	
Other Varieties	6%	
Braeburn	3%	

Basic College Mathematics Sixth Edition, Elayn Martin-Gay   Sec. 9.3

Section 9.3 Mean, Median, and Mode

Complete the outline as you view Lecture Video 9.3. Pause ⏸ the video as needed as you fill in all blanks. Circle your answer to each numbered exercise. Then press Play ▶ to continue listening to the video.

Objective A Find the mean of a list of numbers.

> The mean (average) of a set of number items is the sum of the items divided by the number of items.
> $$\text{mean} = \frac{\text{sum of items}}{\text{number of items}}$$

Find the mean. Round to the nearest tenth.

▶ **Work with me.**

1. 7.6, 8.2, 8.2, 9.6, 5.7, 9.1

⏸ In college, the calculation of a _____ (GPA) is often a weighted _____.

Find the weighted mean, or GPA. Round to the nearest hundredth.

▶ **Work with me.**

2.

Grade	Credit Hours	Point Value of Grade	Point Value · Credit Hours
B	3		
C	3		
A	4		
C	4		
Totals			

Objective B Find the median of a list of numbers.

> The median of a set of numbers in numerical order is the middle number. If the number of items is odd, the median is the middle number. If the number is of items is even, the median is the mean of the two middle numbers.

132

Copyright © 2019 Pearson Education, Inc.

Section 9.3 Mean, Median, and Mode

Find the median.

🔘 **Work with me.**

3. 7.6, 8.2, 8.2, 9.6, 5.7, 9.1

Objective C Find the mode of a list of numbers.

The mode of a set of numbers is the number that occurs most often. (It is possible for a set of numbers to have more than one mode or to have no mode.)

Find the mode.

🔘 **Work with me.**

4. 7.6, 8.2, 8.2, 9.6, 5.7, 9.1

Basic College Mathematics Sixth Edition, Elayn Martin-Gay — Sec. 9.4

Section 9.4 Counting and Introduction to Probability

Complete the outline as you view Lecture Video 9.4. Pause the video as needed as you fill in all blanks. Circle your answer to each numbered exercise. Then press Play to continue listening to the video.

Objective A Use a tree diagram to count outcomes.

> Each chance happening is called an _____. The possible results of an experiment are called _____.

Determine the outcomes of the experiment.

Work with me.

1. Choose a letter in the word MATH, then a number (1, 2, or 3).

Objective B Find the probability of an event.

> _____ is the measure of the chance or likelihood of an event occurring.

The Probability of an Event

$$\text{probability of an event} = \frac{\text{number of ways that the event can occur}}{\text{number of possible outcomes}}$$

Examine the standard six sided die. Determine the probability in each exercise.

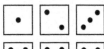
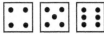

Work with me.

2. If a single die is tossed once, find the probability of a 5.

134
Copyright © 2019 Pearson Education, Inc.

Section 9.4 Counting and Introduction to Probability

> The probability of something impossible to happen is 0.
> The probability of something certain to happen is 1.

Work with me.

3. If a single die is tossed once, find the probability of an even number.

Work with me.

4. If a spinner is spun once, what is the probability of the result of a 2?

Section 10.1 Signed Numbers

Complete the outline as you view Lecture Video 10.1. Pause the video as needed as you fill in all blanks. Circle your answer to each numbered exercise. Then press Play to continue listening to the video.

Objective A Represent real-life situations with signed numbers.

Together, we call positive numbers, zero, and negative numbers the _____ _____.

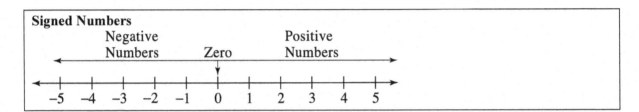

Represent the situation with a signed number.

Work with me.

1. A worker in a silver mine in Nevada works 1235 feet underground.

Objective B Graph signed numbers on a number line.

Graph the set of numbers on a number line.

Work with me.

2. $-3, 0, 4, -1\frac{1}{2}$

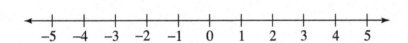

The _____ consist of zero, the natural numbers, and their opposites.

Objective C Compare signed numbers.

On a number line, the number to the left is the smaller number and the number to the right is the greater number.

Basic College Mathematics Sixth Edition, Elayn Martin-Gay   Sec. 10.1

Section 10.1 Signed Numbers

Compare the signed numbers by placing the correct inequality symbol between them.

🔵 **Work with me.**

3. −5   −7

⏸ **Pause and work.**

4. 0   −3

🔵 **Play and check.**

🔵 **Work with me.**

5. $-2\frac{1}{2}$ $-\frac{9}{10}$

Objective D Find the absolute value of a number.

⏸ The _____ _____ of a number is that number's distance from 0 on a number line.

Find the absolute value of each number.

🔵 **Work with me.**

6. $|-8|$

⏸ **Pause and work.**

7. $|0|$

🔵 **Play and check.**

🔵 **Work with me.**

8. $|-8.1|$

Basic College Mathematics Sixth Edition, Elayn Martin-Gay   Sec. 10.1

Section 10.1 Signed Numbers

Objective E **Find the opposite of a number.**

> Two numbers whose distance from 0 is the same, but who lie on opposite sides of 0 are called _____ of each other.

Work with me.

9. The opposite of 5 is ___.

Pause and work.

10. The opposite of −4 is ___.

Play and check.

Work with me.

11. The opposite of $-\dfrac{9}{16}$ is ___.

Opposites
If a is a number, then $-(-a) = a$.

Evaluate each expression.

Work with me.

12. $-(-8)$

Pause and work.

13. $-|-3|$

Play and check.

Basic College Mathematics Sixth Edition, Elayn Martin-Gay   Sec. 10.1

Section 10.1 Signed Numbers

Objective F Read bar graphs containing signed numbers.

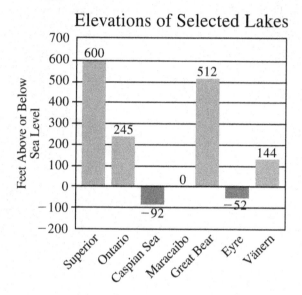

Use the graph above to answer each question.

🔵 **Work with me.**

14. Which lake shown has the lowest elevation?

⏸ **Pause and work.**

15. Which lake shown has the highest elevation?

🔵 Play and check.

Basic College Mathematics Sixth Edition, Elayn Martin-Gay Sec. 10.2

Section 10.2 Adding Signed Numbers

Complete the outline as you view Lecture Video 10.2. Pause the video as needed as you fill in all blanks. Circle your answer to each numbered exercise. Then press Play to continue listening to the video.

Objective A Add signed numbers.

Add the signed numbers, using the number line as necessary.

Work with me.

1. $-1+(-6)$

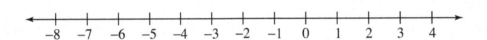

Write the steps used to add two numbers with the same sign.

Adding Two Numbers with the Same Sign
Step 1:
Step 2:

Add the signed numbers.

Work with me.

2. $-6+(-2)$

Section 10.2 Adding Signed Numbers

Add the signed numbers, using the number line as necessary.

▶ **Work with me.**

3. $-4 + 7$

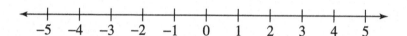

⏸ Write the steps used to add two numbers with different signs.

Adding Two Numbers with Different Signs

Step 1:

Step 2:

Add the signed numbers.

▶ **Work with me.**

4. $6 + (-2)$

⏸ **Pause and work.**

5. $5 + (-9)$

▶ **Play and check.**

▶ **Work with me.**

6. $-10.7 + 15.3$

Section 10.2 Adding Signed Numbers

⏸ **Pause and work.**

7. $-\dfrac{2}{3} + \left(-\dfrac{1}{6}\right)$

▶ **Play and check.**

▶ **Work with me.**

8. $-4 + 2 + (-5)$

Objective B Solve problems by adding signed numbers.

▶ **Work with me.**

9. Suppose a deep-sea diver dives from the surface to 215 feet below the surface. He then dives down 16 more feet. Use positive and negative numbers to represent this situation. Then find the diver's present depth.

Section 10.3 Subtracting Signed Numbers

Complete the outline as you view Lecture Video 10.3. Pause the video as needed as you fill in all blanks. Circle your answer to each numbered exercise. Then press Play to continue listening to the video.

Objective A Subtract signed numbers.

Subtracting Two Numbers
If a and b are numbers, then $a - b = a + (-b)$.

Subtract the signed numbers.

Work with me.

1. $3 - 8$

Pause and work.

2. $-5 - (-8)$

Play and check.

Pause and work.

3. $-14 - 4$

Play and check.

Pause and work.

4. $\dfrac{2}{5} - \dfrac{7}{10}$

Play and check.

Section 10.3 Subtracting Signed Numbers

> Order matters when subtracting.

Pause and work.

5. Subtract 2 from 6.

Play and check.

Pause and work.

6. Subtract -11 from 2.

Play and check.

Objective B Add and subtract signed numbers.

Add and subtract the signed numbers.

Work with me.

7. $-5 - 8 - (-12)$

Objective C Solve problems by subtracting signed numbers.

Solve the problem.

Work with me.

8. The coldest temperature ever recorded on Earth was $-129°F$ in Antarctica. The warmest temperature ever recorded was $134°F$ in Death Valley, California. How many degrees warmer is $134°F$ than $-129°F$?

Basic College Mathematics Sixth Edition, Elayn Martin-Gay   Sec. 10.4

Section 10.4 Multiplying and Dividing Signed Numbers

Complete the outline as you view Lecture Video 10.4. Pause ⏸ the video as needed as you fill in all blanks. Circle your answer to each numbered exercise. Then press Play ▶ to continue listening to the video.

Objective A Multiply signed numbers.

> **Multiplying Numbers**
> The product of two numbers having the same sign is a positive number.
>
> Product of Like Signs
> $\quad (+)(+) = +$
> $\quad (-)(-) = +$
>
> The product of two numbers having different signs is a negative number.
>
> Product of Different Signs
> $\quad (-)(+) = -$
> $\quad (+)(-) = -$

Multiply as indicated.

▶ **Work with me.**

1. $-4(9)$

⏸ **Pause and work.**

2. $-4(4)(-5)$

▶ **Play and check.**

▶ **Work with me.**

3. $\left(-\dfrac{3}{5}\right)\left(-\dfrac{2}{7}\right)$

Basic College Mathematics Sixth Edition, Elayn Martin-Gay   Sec. 10.4

Section 10.4 Multiplying and Dividing Signed Numbers

⏸ **Pause and work.**

4. $(-2)^2$

▶ **Play and check.**

▶ **Work with me.**

5. $(-3)^3$

Objective B Divide signed numbers.

Dividing Numbers
The quotient of two numbers having the same sign is a positive number.

Quotient of Like Signs
$$\frac{(+)}{(+)} = + \qquad \frac{(-)}{(-)} = +$$

The quotient of two numbers having different signs is a negative number.

Quotient of Different Signs
$$\frac{(+)}{(-)} = - \qquad \frac{(-)}{(+)} = -$$

Divide as indicated.

▶ **Work with me.**

6. $-24 \div 6$

⏸ **Pause and work.**

7. $\dfrac{-88}{-11}$

▶ **Play and check.**

146

Section 10.4 Multiplying and Dividing Signed Numbers

Work with me.

8. $\dfrac{0}{14}$

Pause and work.

9. $\dfrac{-13}{0}$

Play and check.

Work with me.

10. $\dfrac{-120}{0.4}$

The fraction $-\dfrac{a}{b} = \dfrac{-a}{b} = \dfrac{a}{-b}$

Objective C Solve problems by multiplying and dividing signed numbers.

Work with me.

11. A football team lost four yards on each of three consecutive plays. Represent the total loss as a product of signed numbers and find the total loss.

Basic College Mathematics Sixth Edition, Elayn Martin-Gay Sec. 10.5

Section 10.5 Order of Operations

Complete the outline as you view Lecture Video 10.5. Pause the video as needed as you fill in all blanks. Circle your answer to each numbered exercise. Then press Play to continue listening to the video.

Objective A Simplify expressions by using the order of operations.

Write the steps used in order of operations.

Order of Operations
Step 1:
Step 2:
Step 3:
Step 4:

Evaluate each expression.

Work with me.

1. $-1(-2)+1$

Pause and work.

2. $\dfrac{4}{9}\left(\dfrac{2}{10}-\dfrac{7}{10}\right)$

Play and check.

148

Basic College Mathematics Sixth Edition, Elayn Martin-Gay Sec. 10.5

Section 10.5 Order of Operations

Work with me.

3. $\dfrac{24}{10+(-4)}$

Pause and work.

4. $3^3 - 12$

Play and check.

Work with me.

5. $|8-24| \cdot (-2) \div (-2)$

Pause and work.

6. $2 - 7 \cdot 6 - 19$

Play and check.

Work with me.

7. $\dfrac{(-7)(-3) - 4(3)}{3[7 \div (3-10)]}$

149

Basic College Mathematics Sixth Edition, Elayn Martin-Gay Sec. 10.5

Section 10.5 Order of Operations

Objective B Find the average of a list of numbers.

Average of a List of Numbers

$$\text{average} = \frac{\text{sum of numbers}}{\text{number of numbers}}$$

Find the average of the list of numbers.

Work with me.

8. $-10, 8, -4, 2, 7, -5, -12$

Basic College Mathematics Sixth Edition, Elayn Martin-Gay   Sec. 11.1

Section 11.1 Introduction to Variables

Complete the outline as you view Lecture Video 11.1. Pause the video as needed as you fill in all blanks. Circle your answer to each numbered exercise. Then press Play to continue listening to the video.

Objective A Evaluate algebraic expressions for given replacement values for the variables.

- A letter used to represent a number is called a _____.

- An _____ has letters (variables) and numbers combined with operation symbols.

- _____ means to find the value of.

Evaluate.

Work with me.

1. $\dfrac{x+2y}{2z}$ when $x = -2$, $y = 5$, $z = -3$

Objective B Use properties of numbers to combine like terms.

- The _____ of an expression are the addends of the expression.

- A _____ _____ contains a variable.

- A _____ _____ (or sometimes just a constant) contains a number only.

- The _____ _____ of a term is the numerical factor.

- _____ _____ have the same variable factors, but the numerical factors may be different.

Section 11.1 Introduction to Variables

> **Distributive Property**
> If a, b, and c are numbers, then $ac + bc = (a+b)c$
> Also, $ac - bc = (a-b)c$

Combine like terms.

🔘 **Work with me.**

2. $3x + 5x$

Simplify.

🔘 **Work with me.**

3. $3x + 7 - x - 14$

⏸ **Pause and work.**

4. $-5m - 2.3m + 11 + 2.5m - 15.1$

🔘 **Play and check.**

Objective C Use properties of numbers to multiply expressions.

Multiply.

🔘 **Work with me.**

5. $-2(11y)$

Basic College Mathematics Sixth Edition, Elayn Martin-Gay   Sec. 11.1

Section 11.1 Introduction to Variables

⏸ **Pause and work.**

6. $-4(3x+7)$

▶ Play and check.

Objective D **Simplify expressions by multiplying and then combining like terms.**

Simplify.

▶ **Work with me.**

7. $3+6(w+2)+w$

Objective E **Find the perimeter and area of figures.**

⏸ _____ measures the distance around a polygon. It is measured in units.

▶ **Work with me.**

8. Determine the perimeter and area of the rectangle.

   $(3y + 1)$ miles

   20 miles

⏸ _____ measures the surface of a region. It is measured in square units.

Basic College Mathematics Sixth Edition, Elayn Martin-Gay — Sec. 11.2

Section 11.2 Solving Equations: The Addition Property

Complete the outline as you view Lecture Video 11.2. Pause ⏸ the video as needed as you fill in all blanks. Circle your answer to each numbered exercise. Then press Play ▶ to continue listening to the video.

Objective A Determine whether a given number is a solution of an equation.

> Equation: expression = expression

⏸ A _____ of an equation is a value for the variable that makes the equation a true statement.

▶ **Work with me.**

1. Is -5 a solution of $x + 12 = 17$?

Objective B Use the addition property of equality to solve equations.

> **Addition Property of Equality**
> Let a, b, and c represent numbers. Then
>
> $a = b$ and $a + c = b + c$ are equivalent equations.
>
> Also, $a = b$ and $a - c = b - c$ are equivalent equations.

Solve.

▶ **Work with me.**

2. $a + 5 = 23$

⏸ **Pause and work.**

3. $7 = y - 2$

▶ **Play and check.**

Section 11.2 Solving Equations: The Addition Property

▶ **Work with me.**

4. $y - \dfrac{3}{4} = -\dfrac{5}{8}$

⏸ **Pause and work.**

5. $-2 - 3 = -4 + x$

▶ **Play and check.**

▶ **Work with me.**

6. $7x + 14 - 6x = -4 + (-10)$

▶ **Work with me.**

7. $2(5x - 3) = 11x$

Basic College Mathematics Sixth Edition, Elayn Martin-Gay   Sec. 11.3

Section 11.3 Solving Equations: The Multiplication Property

Complete the outline as you view Lecture Video 11.3. Pause ⏸ the video as needed as you fill in all blanks. Circle your answer to each numbered exercise. Then press Play ▶ to continue listening to the video.

Objective A Use the multiplication property to solve equations.

Multiplication Property of Equality
Let a, b, and c represent numbers and let $c \neq 0$. Then

$a = b$ Also, $a = b$

and $a \cdot c = b \cdot c$ and $\dfrac{a}{c} = \dfrac{b}{c}$

are equivalent equations.   are equivalent equations.

Solve.

▶ **Work with me.**

1. $-3z = 12$

Divide both sides by the coefficient of the variable.

▶ **Work with me.**

2. $-0.3x = -15$

⏸ **Pause and work.**

3. $\dfrac{8}{5}t = -\dfrac{3}{8}$

▶ **Play and check.**

156

Section 11.3 Solving Equations: The Multiplication Property

Work with me.

4. $5 - 5 = 2x + 7x$

Pause and work.

5. $23x - 25x = 7 - 9$

Play and check.

Pause and work.

6. $18 - 11 = \dfrac{x}{5}$

Play and check.

Basic College Mathematics Sixth Edition, Elayn Martin-Gay   Sec. 11.4

Section 11.4 Solving Equations Using Addition and Multiplication Properties

Complete the outline as you view Lecture Video 11.4. Pause the video as needed as you fill in all blanks. Circle your answer to each numbered exercise. Then press Play to continue listening to the video.

Objective A Solve linear equations using the addition and multiplication properties.

Solve.

Work with me.

1. $-7c + 1 = -20$

Multiplication Property of Equality - Divide _____ sides by the nonzero coefficient of the variable.

A _____ of an equation is a value of the variable that makes the equation a true statement.

Solve.

Work with me.

2. $10x + 15 = 6x + 3$

Always check your proposed solution in the original equation.

Objective B Solve linear equations containing parentheses.

Solve.

Work with me.

3. $3(5c - 1) - 2 = 13c + 3$

Section 11.4 Solving Equations Using Addition and Multiplication Properties

Write the steps used to solve an equation.

Steps for Solving an Equation

Step 1:

Step 2:

Step 3:

Step 4:

Step 5:

Solve.

Pause and work.

4. $3(x-1)-12=0$

Play and check.

Objective C Write numerical sentences as equations.

Work with me.

Translate.

5. The product of −5 and −29 gives 145.

Pause and work.

6. Three times the difference of −14 and 2 amounts to −48.

Section 11.5 Equations and Problem Solving

Complete the outline as you view Lecture Video 11.5. Pause the video as needed as you fill in all blanks. Circle your answer to each numbered exercise. Then press Play to continue listening to the video.

Objective A Write phrases as algebraic expressions.

Addition	Subtraction	Multiplication	Division	Equality
sum	difference	product	quotient	equals
plus	minus	times	divided by	gives
added to	subtract from	multiply	into	is/was
more than	less than	twice	per	yields
increased by	decreased by	of		amounts to
total	less	double		is equals to

Translate.

Work with me.

1. The sum of a number and 5.

Pause and work.

2. Twenty decreased by a number.

Play and check.

Objective B Write sentences as equations.

Translate.

Work with me.

3. Twice a number gives 108.

Basic College Mathematics Sixth Edition, Elayn Martin-Gay   Sec. 11.5

Section 11.5 Equations and Problem Solving

Pause and work.

4. A number subtracted from −20 amounts to 104.

Play and check.

Objective C Use problem-solving steps to solve problems.

Write the steps used to solve application problems.

Problem-Solving Steps
Step 1:
Step 2:
Step 3:
Step 4:

Determine the unknown number.

Work with me.

5. Three times a number, added to 9 is 33.

Section 11.5 Equations and Problem Solving

Work with me.

6. Three times the difference of some number and 5 amounts to the quotient 108 and 12.

Solve the application.

Work with me.

7. A falcon, when diving, can travel five times as fast as a pheasant's top speed. If the total speed for these two birds is 222 miles per hour, find the fastest speed of the falcon and the fastest speed of the pheasant.